COLLECTIVE AQUIFER GOVERNANCE

Current models of groundwater governance focus principally on the allocation of water, rather than taking a holistic approach incorporating valuable storage space in the aquifer, as well as the transformative changes in managed recharge of manufactured water, stormwater, and carbon. Effective implementation of a more modern approach now calls for a rethink of both scale and jurisdictional boundaries. This involves linking public and private aspects of water quantity, water quality, geothermal regulation, property rights, subsurface storage rights, water marketing, water banking, legal jurisdictions, and other components into a single governance document. This style of agreement stands in contrast to the siloed approach currently applied to aquifer resources. Using case studies, and an activity inspired by gaming concepts to explore the incentives and challenges to aquifer governance approaches, this book demonstrates how application of the principles of unitization agreements to aquifers could provide a new approach to aquifer governance models.

JAKOB WILEY is an attorney with substantial expertise in water law, water governance organizations, and local and state government law. His research focuses on public and private water organizations and applied legal theory. He has published in several journals and newsletters, including the *Kansas Journal of Law & Public Policy* (2020).

TODD JARVIS is an engineering geologist, water right examiner, and mediator who works at the Institute for Water and Watersheds at Oregon State University. He teaches Environmental Conflict Resolution at the University of Oregon Law School and has more than thirty-five years' experience working throughout the United States and internationally.

COLLECTIVE AQUIFER GOVERNANCE

Dispute Prevention for Groundwater and Aquifers through Unitization

JAKOB WILEY

City of Oregon City

TODD JARVIS

Oregon State University

CAMBRIDGE
UNIVERSITY PRESS

CAMBRIDGE
UNIVERSITY PRESS

University Printing House, Cambridge CB2 8BS, United Kingdom

One Liberty Plaza, 20th Floor, New York, NY 10006, USA

477 Williamstown Road, Port Melbourne, VIC 3207, Australia

314–321, 3rd Floor, Plot 3, Splendor Forum, Jasola District Centre, New Delhi – 110025, India

103 Penang Road, #05-06/07, Visioncrest Commercial, Singapore 238467

Cambridge University Press is part of the University of Cambridge.

It furthers the University's mission by disseminating knowledge in the pursuit of
education, learning, and research at the highest international levels of excellence.

www.cambridge.org
Information on this title: www.cambridge.org/9781107172081
DOI: 10.1017/9781316771365

© Jakob Wiley and Todd Jarvis 2022

First published 2022

A catalogue record for this publication is available from the British Library.

Library of Congress Cataloging-in-Publication Data
Names: Wiley, Jakob, author. | Jarvis, W. Todd, author.
Title: Collective aquifer governance : dispute prevention for groundwater and aquifers through unitization /
Todd Jarvis, Jakob Wiley
Description: New York, NY, USA : Cambridge University Press, 2022. | Includes bibliographical
references and index.
Identifiers: LCCN 2021030242 (print) | LCCN 2021030243 (ebook) | ISBN 9781107172081 (hardback) |
ISBN 9781316771365 (epub)
Subjects: LCSH: Aquifers. | Groundwater. | Dispute resolution (Law).
Classification: LCC GB1005 .J37 2022 (print) | LCC GB1005 (ebook) | DDC 333.91/04–dc23
LC record available at https://lccn.loc.gov/2021030242
LC ebook record available at https://lccn.loc.gov/2021030243

ISBN 978-1-107-17208-1 Hardback

Contents

Figures

Tables

Acknowledgments

The authors acknowledge John W. Jarvis, petroleum landman who developed many unitization and joint operating agreements over a thirty-year career in the oil industry. John provided important guidance on comparing and contrasting unitization concepts between oil and water and started our research interests in applying unitization to aquifers in 2008. Christopher Adams, the commercial artist of Corvidopolis, Corvallis, Oregon, drew the "lawgineer" diagrams for us based on written description from the 1930s and today. Caryn M. Davis of Cascadia Editing, Philomath, Oregon, completed expert manuscript editing and formatting. Historian Kimberly S. Jensen provided important references to the history of conservation and Gifford Pinchot.

Jakob Wiley would like to acknowledge Todd Jarvis for being a great mentor and trusting him with the research into the fascinating topic discussed in this book. Todd's infinite patience and vast wisdom were critical to Jakob's success. Jakob would also like to acknowledge his parents, Tom and Meridee Wiley, for their sacrifices and support during the process of writing this book. Jakob would also like to thank his cats, Izzy and Kirby, for their endless encouragement.

Thanks to all for helping us reach the finish line.

Abbreviations

ARUs	aquifer recharge units
ASR	aquifer storage and recovery
CAGAs	collective aquifer governance agreements
CERCLA	Comprehensive Environmental Response, Compensation, and Liability Act
cfs	cubic feet per second
CPROs	common-pool resource organizations
CVWD	Coachella Valley Water District
DLoTA	Draft Articles of the Law of Transboundary Aquifers
DWA	Desert Water Agency
EER	estimated economic recovery
EPA	Environmental Protection Agency
ER	estimated recovery
EVWUA	Escalante Valley Water Users Association
GGRETA	Governance of Groundwater Resources in Transboundary Aquifers
HIIP	hydrocarbons initially in place
HVAC	Greater Harney Valley Groundwater Area of Concern
IGRAC	International Groundwater Assessment Centre
IGUCA	intensive groundwater use control area
IHPV	initial hydrocarbon pore volume
IMAR	incentivized managed aquifer recharge
IWRM	integrated water resources management
LEMA	local enhanced management areas
MAR	managed aquifer recharge
MMSIC	Model Memphis Sands Interstate Compact
MWD	Metropolitan Water District
ORS	Oregon Revised Statute
OWRD	Oregon Water Resources Department

PGIS   participatory geographic information system
PIA   practicably irrigable acres
SGAs   sustainable groundwater agencies
SGMA   Sustainable Groundwater Management Act
SGPs   sustainable groundwater plans
SWP   California State Water Project
TDS   total dissolved solids
UNESCO   United Nations Educational, Scientific and Cultural
              Organization

1

Introduction

Conflicts, big or little, should never happen; someone always gets
licked – 99 times of out of 100, yes, 999 times out of 1,000, both sides
get licked.

—*H. L. Doherty, 1925 (quoted in Campbell, 1925)*

Groundwater issues are becoming increasingly complex and large in geographic
extent, almost too large for the traditional centralized management by governmental agencies. Consider, for example, that the global groundwater use is 800–1,000
km^3 per year (200–260 trillion gallons per year or 650–810 million acre-feet per
year). Fifty percent of annual global groundwater use is for agriculture by India,
Pakistan, Bangladesh, and China (Margat & van der Gun, 2013). In the state of
Oregon where we live, many rivers, such as the Metolius, Deschutes, Klamath, and
even the Willamette River, the twenty-third largest river by discharge in the United
States, are composed of significant percentages of discharging groundwater.
Adjacent aquifer development, coupled with the hydraulic connection to rivers, is
fertile ground for domestic and international disputes over compacts, treaties, and
less formal agreements. The term "adjacent" means hydraulically connected to a
surface-water or groundwater body, either in the traditional physical sense or in the
less traditional legal sense. Disputes over *adjacent* oceans started the interest in
public entrepreneurship and groundwater management, initially addressing seawater intrusion in California and progressing to the relocation of Jakarta, Indonesia,
due to land subsidence – caused by the collapse of fine-grained confining layers in
interbedded aquifer systems or closed fractures in dewatered fractured-rock aquifer
systems – associated with groundwater pumping. And *adjacent* arsenic has created
a crisis in drinking water derived from pumped groundwater for tens of millions in
Bangladesh, as well as leaching from the aquifers to the waters used for managed
or artificial recharge (water intentionally placed in an aquifer to restore the water
table) to the groundwater basins in southern California.

1.1 The Race to the Bottom

Depending on whom you ask, the number of wells and groundwater extraction mechanisms in the world approaches tens to hundreds of millions. We compiled a listing of readily available statistics on well numbers by country (Table 1.1). Some researchers indicate that the reported number of wells in China is low and may be increasing rapidly due to the change from a collective agricultural society to a more capitalist and industrialist society. The statistics reported for the United States (14 million) and Canada (1 million) are from groundwater industry databases; Perrone and Jasechko (2019) substantiate the US records by inventorying 11.8 million wells, and they suggest that drilling deeper is not a sustainable solution for groundwater-dependent societies. For comparison, easy-to-access industry and government oil and gas databases report 1.7 million active wells in the United States in 2019, with tens of thousands of new wells drilled each year – compared with millions of groundwater wells. In other words, there is ample opportunity for conflicts both small and very large as the race to access groundwater, the world's largest source of freshwater, continues unabated.

Table 1.1. *Estimated number of wells and groundwater extraction mechanisms in different countries*

Country	Wells
India	19–26 million
United States	12–15 million
China	3.4–3.5 million
Canada	1 million
Germany	500,000
South Africa	500,000
Iran	500,000
Pakistan	500,000
Abu Dhabi	130,000
Denmark	75,000
Mexico	70,000
Taiwan	37,000
Mongolia	27,000
Ireland	10,000
Malta	10,000
Costa Rica	5,000

1.2 The Slow Pace of Agreeing to Disagree

In the current era of groundwater depletion with simultaneous irreversible damage to aquifer storage, new instruments of groundwater governance must focus not only on process equity and outcome equity but also on aquifer governance – what to do to preserve and reuse the storage characteristics of the container holding the water. While the International Groundwater Assessment Centre in the Netherlands reports there are approximately 592 "shared" aquifers, Burchi (2018) reports there are less than ten agreements on record regarding aquifers that straddle international boundary lines. The agreements for transboundary aquifers in chronological order are as follows:

- Agreement first made in 1977, then renegotiated in 2007 by France and Switzerland on the Genevese Aquifer
- Three agreements on the Nubian Sandstone Aquifer System, made in 1992 and 2000 by Chad, Egypt, Libya, and Sudan
- Agreement for the establishment of a trilateral consultative arrangement for the North-Western Sahara Aquifer System, made by Algeria, Libya, and Tunisia in the period 2002–2008
- Agreement (technically, a memorandum of understanding) made in 2009 by Mali, Niger, and Nigeria for the establishment of a trilateral consultative arrangement for the Iullemeden Aquifer System (IAS). The agreement is not in force yet. A memorandum of understanding was agreed to in 2014 by Algeria, Benin, Burkina Faso, Mali, Mauritania, Niger, and Nigeria for the Iullemeden and Taoudeni/Tanezrouft Aquifer Systems (ITAS), for the establishment of a comparable multipartite Consultative Mechanism for the ITAS. This later agreement is also not in force yet.
- Agreement on the Guarani Aquifer made in 2010 by Argentina, Brazil, Paraguay, and Uruguay
- Agreement on the Al-Sag/Al-Disi Aquifer made in 2015 by Jordan and Saudi Arabia
- ORASECOM Resolution on Stampriet Transboundary Aquifer System Multi-Country Cooperation Mechanism, which will continue the joint study and characterization, generate flow of data feeding a borehole database and numerical model, and report on activities at each meeting of the Groundwater Hydrology Committee, made in 2017 by Namibia, Botswana, and South Africa.

However, none of these agreements address the aquifer storage per se; rather, they address the groundwater stored in the aquifer.

At the international level, the Convention on the Law of Non-Navigational Uses of International Watercourses (UN Watercourses Convention) entered into force in August 2014 when Vietnam became the thirty-fifth country to accede to the convention. Yet, groundwater and aquifers are addressed in an overly simplified approach, one that specifically focuses on the hydraulic connection to a watercourse. "Fossil" aquifers, or aquifers storing nonrenewable groundwater, and coastal aquifers are not specifically addressed in the UN Watercourses Convention. The Draft Articles of the Law of Transboundary Aquifers (DLoTA), which is now annexed to a United Nations General Assembly Resolution, details the use of aquifers that extends beyond just groundwater: "utilization of transboundary aquifers or aquifer system includes extraction of water, heat and minerals, and storage and disposal of any substance" (Eckstein, 2017, p. 145). This is an important acknowledgment that the available aquifer storage is also an important transboundary resource that must be collectively managed beyond considering the aquifer as just a container for storing recoverable groundwater.

1.3 The Aquifer Storage Conundrum

Not all aquifer storage is created equal. Shah (2009) describes the notion of aquifer communities, where there is the recognition that aquifer storage conditions affect human behavior and the institutional response of groundwater users. Shah defines an aquifer community as users in a locality who are aware of their interdependence in the development or conservation of a common aquifer or a portion thereof that shapes their individual or collective behavior. He delineates five situations in India that depend on the aquifer's characteristics. Note that the governance models in these situations are dependent on the storage characteristics of the developed aquifer and are not simply exploring how to share groundwater:

- Sand and gravel aquifer, high storage with good recharge – no opportunity for aquifer governance because the impact on a typical user is insignificant and there is no need for aquifer communities. Good governance is planning the expansion of groundwater use for maximum social welfare and poverty reduction.
- Sand and gravel aquifer, high storage with limited to no recharge – no to low opportunity for aquifer governance, as long as the users can chase a falling water level and do not recognize the interdependence between groundwater users. Good governance is improved conjunctive management of groundwater, surface water, and rainwater.
- Hard rock aquifer, low storage with some recharge – some opportunity for aquifer governance as the interdependence of users is known, leading to a rise in numerous aquifer communities as there is a near zero-sum game of competitive deepening among well owners, but also efficient water use. Good governance is

mobilizing water harvesting and decentralized recharge, in addition to local demand management.

- Hard rock aquifer with low storage or alluvial aquifers with confining layer – high opportunity for aquifer communities, as there is strong recognition of interdependence. Good governance is institutionalizing the positive experiences and building upon them, in addition to demand management.
- Arid or coastal aquifers susceptible to rapidly changing water quality – no opportunity for aquifer governance because the water quantity impact on a typical user is insignificant, but the water quality impact is swift. Aquifer communities are incapable of reviving the aquifer and they exit the area. Good governance is large-scale public intervention on supply and demand. The goal is to improve management of surface water.

As proposed by Shah (2009) for India, the fragmented nature of water and land use at the individual state level in the United States, due in part to the lack of integration between land use and water laws, is leading to a new paradigm in water planning and management that focuses on a "bottom-up" approach instead of the traditional "top-down" approach. Different "scales" of groundwater governance and management have evolved since 2000. For example, concurrency laws for proposed land use have evolved to address groundwater recoverability and aquifer mechanics, effectively redetermining recoverability and repurposing storativity. These changes came about due to highly variable well yields unrelated to groundwater recharge or depletion; rather, the well yields were related to damaged and lost aquifer storage.

The limits of the current groundwater and aquifer management models must change for the future. More attention must focus on the value of the storage in the aquifer, especially in light of the transformative changes in managed recharge of manufactured water and stormwater, rather than on the allocation of water. To do this effectively, it is time to rethink scale and jurisdictional boundary spanning.

1.4 Subsurface Collective Action through Unitization?

All of these situations could easily be brought under the notion of "groundwater and aquifer governance by contract." Our research on groundwater governance has focused more on the notion of unitization of aquifers as a holistic approach to groundwater governance. Unitization was the answer for the oil industry's tragedy of the commons over a century ago. Unitization is the well-known joint operation of oil or gas reservoirs by all of the owners of rights in the separate tracts overlying the reservoirs that has been in practice for over 100 years. "Pooling" is sometimes referred to as unitization. Table 1.2 provides a comparison between unitization that

Table 1.2. *Comparison between unitization in the oil industry and proposed approach for aquifers*

Stage of field life	Unitization status (Worthington, 2011)	Petroleum class	Proposed aquifer unitization
Collaboration			Principled collaboration compact
Exploration		Prospective	Pre-unit agreement based on voluntary geographic unit
Discovery		Contingent	
Appraisal	Pre-unit agreement		Pre-unit agreement based on voluntary geologic unit
Commerciality		Reserves (UD)	
Development	Unitization and operating agreement		Unitization and operating agreement
Production	Redetermination(s)	Reserves (D)	Redetermination and compulsory/conservation units
Abandonment			Redetermination(s) of new use of aquifer

has been applied over the past 60–70 years in the oil industry and our proposed approach for aquifers.

Unitization, as employed in the oil industry, is designed to be collectively beneficial. It is practiced in 38 states and 13 countries, and most recently, it is the proposed approach for sharing transboundary hydrocarbon resources in the Gulf of Mexico, as outlined in the US–Mexico Transboundary Hydrocarbons Agreement of 2012 (Hagerty & Uzel, 2013). But the practice of unitization to the subsurface is not limited to oil and gas. This is an important development, given the recent news of huge sub-seabed aquifers, which begs the question as to how these will best be governed: as part of the global commons, through the Law of the Sea or the Law of the Hidden Sea, or through some form of contract or operating agreement, as described by Martin-Nagle (2016, 2020).

Many lawyers and the public conflate the unitization of aquifers with privatizing groundwater, or believe that it could not be applied in the western United States. However, careful examination of how unitization is applied focuses more on the "container" – the reservoir or aquifer – than on what is stored in the container. Far from overturning established water law, unitization merely harmonizes water use within the current water rights system. Obviously, water management models vary from state to state in the United States, and yet, as discussed later in this book, vestiges of unitization and collective action for aquifers are occurring regardless of whether groundwater rights and associated aquifer storage space are allocated through prior appropriation (Utah, Idaho); reasonable use/correlative rights (California); or absolute ownership (Texas). We are also seeing the issue of aquifer storage being debated as part of Federal Reserved Rights, which is being asked of the United States Supreme Court (*Desert Water Authority & Coachella Valley Water District* v. *Agua Caliente Band of Cahuilla Indians*).

With the increasing interest in groundwater exploration by entrepreneurs such as the Pacific Aquifer Exploration Syndicate, formed by the mining company Canadian International Minerals Inc., there will be increased interest in unitization as applied to aquifers through the creation of "voluntary units," or agreements among interested parties that can be undertaken for exploration; this is a common practice in the oil industry. We also suggest that unitization might be applied to repurposing the storage space in depleted aquifers, or that the unitization of contaminated groundwater may provide new opportunities to remediate and market the previously unusable water and storage. As discussed in subsequent chapters, the collective-action approach to subsurface governance premised on the concepts of unitization has historical precedence for both private and public enterprises.

1.5 Myth Busting

When we prepared this book proposal, we received many excellent reviewer comments for improving the text. Yet we also received many comments that suggested a misunderstanding of unitization both in the context of oil and gas, as well as in how it might be applied to groundwater and aquifers. We felt it important to address a couple of these issues at the beginning of the book and hope that the others are addressed elsewhere in the text.

1.5.1 Unitization Is Based on the Design Principles of Common Pool Resources

We have heard this assessment regarding Henry Doherty's (1924) concept of oil and gas unitization: that it sounds like it takes after the design principles developed by Elinor Ostrom (1990) for common pool resources. We reproduce a preliminary summary of the two governance frameworks, as described by Jarvis (2011), in Table 1.3. On the basis of history alone, the two fields of study were introduced decades apart, thus dispelling the myth that unitization principles were adapted from *Governing the Commons* (Ostrom, 1990).

But the comparison also introduced a new layer of inquiry: Why was there a general lack of recognition of unitization by Ostrom's body of work on groundwater and the utilization of aquifers? A deeper dive into the historical references determined that Ostrom's remarkable and highly readable 609-page doctoral dissertation, discussed in later sections, focused on aquifer conservation by *public*

Table 1.3. *Design attributes and principles of unitization versus common pool resources*

Principle or attribute	Unitization	Common pool resources
Conceptual development	1890s–1930s	1960s–1990s
Organizations Boundaries	Private Enterprise • Voluntary units • Compulsory/conservation units • Geographic units • Geologic units	Public Enterprise Clearly define boundaries for the user pool and the resource domain
Rules	• Pre-unit agreements at appraisal • Unitization agreement at pre-development • Redetermination during development	Appropriation rules developed for local conditions and provisional rules developed for resource maintenance
Collective action	• Collectively beneficial • Allows sharing of development infrastructure • Avoids unnecessary wells and infrastructure occurring under the competitive rule of capture	Collective choice arrangements developed by the resource users
Monitoring	• Uses pressure maintenance on the reservoir • Uses best technical or engineering information • Provides foundation to carry out a secondary recovery program	Monitoring programs developed for the resource
Sanctions	Gives all owners of rights in the common reservoir a fair share of the production	Graduated sanctions developed for "violators" of the rules
Dispute resolution	• Pre-unit agreement • Industry-standard agreements • Redetermination process	Conflict management schemes developed
Rights of regimes	Can be developed through voluntary or government-mandated compulsory action	Rights of organized environmental regimes respected by external authorities
Administration	• Voluntary to compulsory • Other alternatives (e.g., sole development, partitioned development, fixed equity, buy-out, or asset swaps)	Nested enterprises used to administer management

Modified after Jarvis (2011)

enterprises in southern California (Ostrom, 1965). Ostrom described public enterprises as providing public goods and services and notes that "economists have long been concerned with entrepreneurship, but have largely confined their analysis of entrepreneurship to the private market economy" (p. 90).

We examine in later sections how Doherty's work focused on oil and gas reservoir conservation through the lens of *private enterprise* and related market economy. At the time of Ostrom's PhD work, the state of California already had in place one of the first unitization agreements in the United States: the 1931 agreement at the Kettleman Hills field, which was largely owned by the federal government. California also had limited compulsory unitization laws, largely directed to areas where land subsidence had occurred due to oil and gas withdrawals. Compulsory unitization was under consideration in California in the mid- to late 1950s but was not warmly received because of the perception that it did not safeguard the public interest from potential monopolies connected to unitization (Weaver, 1986). The debate over unitization at the time of Ostrom's research may be one of the reasons why there was no apparent connection between the design principles of unitization of oil reservoirs and aquifers in California.

1.5.2 Exploitation and Economics

One could argue that unitization has traditionally focused on exploiting oil and gas as quickly as possible, and that such an approach will not work for groundwater. For example, Blumberg and Collins (2016, p. 74) posit that "from a resource conservation perspective, the most important difference between pooling of water interests and pooling of oil and gas interests is that oil and gas production expressly seeks to extract as much of the resource as economically possible."

We are not certain where the confusion between oil and gas production and water production began, so we completed a deep dive into the history of unitization and pooling. Doherty's original plan on unitization was to recruit the federal government to *conserve* the nation's natural resources, and in the case of oil particularly, because

practically every evil on the oil business, and everything about which the public complain, is due to the fact that oil does not follow the usual law of property rights but belongs to the man who can capture it. Other mineral products are located and blocked out but are only taken from their ground reserves as the market needs them. The discovery of an oil pool means that every landowner or lessee can take as much oil from this *common pool* as he can get, and there is always a frenzied scramble to bring the oil to the surface before somebody else can get it regardless of whether the market needs it or not.

(Doherty Letter to President Calvin Coolidge dated
August 11, 1924, in Hardwicke (1948, p. 181))

Doherty summarized his observations in a letter as follows:

(1) That our present methods are viciously wasteful in every way; (2) That the first step in conservation must be to provide that ownership shall be determined otherwise than by capture; (3) That if our laws are changed to make oil and gas conform to the laws governing all other property that its division among different land-owners can be made with a greater relative degree of equity than now prevails; (4) That if we can develop our pools without undue haste we can recover at least double as much oil as we do now; among others.

(Doherty Letter to President Calvin Coolidge dated August 11, 1924, in Hardwicke (1948, p. 181))

Originally antagonistic toward Doherty's plan, geologists and engineers J. M. Lovejoy, E. Degolyer, and W. A. Sinsheimer later came around to Doherty's way of thinking by stating that with the "realization of the fact that each individual oil pool is a *mechanical unit* and should be operated as such regardless of divisions of ownership, [Doherty] became the great protagonist of unit operation." (Preface to the Henry L. Doherty Memorial Volume on *Petroleum Conservation*, July 20, 1948, by J. M. Lovejoy, E. DeGolyer, and W. A. Sinsheimer in Hardwicke, 1948, p. 258, emphasis added). In other words, Doherty became the first to recognize that the oil pool and the reservoir rock operated in a symbiotic relationship: Damage to one common pool resource (oil reservoir rock) would preclude the use of the other common pool resource (recovered oil).

Unitization of oil and gas reservoirs focuses on preserving the gas drive of the reservoir and precluding the intrusion of water in an otherwise hydrocarbon wet reservoir. By comparison, development of groundwater from a confined aquifer releases water from aquifer storage by two mechanisms: compression of the aquifer and expansion of water. However, pumping of groundwater could also drown an aquifer with poor quality water from other aquifers or seawater intrusion.

The concept of unitization applied to both oil reservoirs and aquifers is to preserve the storage characteristics of the reservoir and aquifer. For example, Ostrom (1965) documented wasting groundwater storage through seawater intrusion as an undesirable result in southern California. Damaged and lost aquifer storage due to groundwater pumping eventually led to widespread drying of wells tapped from compressible fractured rocks in eastern Utah in the early 2000s (Jarvis, 2011). As discussed more fully in later sections, the Sustainable Groundwater Management Act of 2014 identified other undesirable results as part of addressing groundwater depletion in California (Table 1.4).

While hydrocarbons are usually regarded as private property, the production of oil and gas has always been treated as affecting the public interest because it has beneficial use to both the *private* and *public* welfare and is an important revenue stream for many states and countries (Jarvis, 2011). As a consequence, we consider the economic arguments against unitization of aquifers to be misinformed; rather,

Table 1.4. *Summary of waste and undesirable results for select subsurface resources*

Wasteful haste in oil reservoirs (Mid-Continent Oil and Gas Association, 1930)	Undesirable results for groundwater and aquifers (SGMA of 2014)
Underground waste, including the operation of wells in excess of their maximum efficient rate	Chronic lowering of groundwater levels
Operating an oil well with an inefficient gas–oil ratio	Significant and unreasonable reduction of groundwater storage
Drowning with formation water of any stratum capable of producing oil or gas in paying quantities	Significant and unreasonable seawater intrusion
Allowing the escape of oil or gas from one stratum to another	Significant and unreasonable degraded water quality, including the migration of contaminant plumes that impair water supplies
Surface waste or loss	Depletions of interconnected surface water
Permitting any gas well to burn wastefully, permitting the escape of casinghead gas from an oil well, or use of gas for wasteful purposes	Significant and unreasonable land subsidence that substantially interferes with surface land uses

the parallels between the historic concepts supporting the unitization of oil and gas reservoirs are on par with the modern notions of undesirable results associated with governing and managing aquifers.

1.5.3 Water Is Different?

We often hear that the differences between water and oil and gas make the concept unworkable. Specifically, that the laws applicable to water are so different from those that apply to oil and gas that unitization of aquifers would not be possible. From today's perspective, the regulation of these two kinds of resources might appear wildly divergent. However, each resource's set of rules has a common origin in legal theory.

The development of rights, laws, and regulations for these two resources followed parallel tracks. The rights associated with both resources have origins in the application of Roman legal principles of *ferae naturae*, or rule of capture (Weaver, 1986). Regulation of these resources began in an effort to reduce conflicts between codevelopers and assign limits to assigned property rights. As the resources became more fully developed, concerns over waste of the whole resource became the focus of legal attention rather than individual conflict. In the

Table 1.5. *Oil and gas regulation compared to groundwater regulation through time*

Issue	Oil and gas	Groundwater
Establishment of a System or Rights of Access	1800s: • Ferae naturae doctrine and the rule of capture (Weaver, 1986)	1800s: • Ferae naturae doctrine and related legal theories like prior appropriation and the rule of capture (Getzler, 2004)
Conflicts between adjacent resource users	Early 1900s: • Courts augment the rule of capture with the concept of "correlative rights" (early 1900s) (Kramer & Anderson, 2005)	Late 1800s through early 1900s: • Riparian and prior appropriation doctrines emerge to address conflicts (Barbanell, 2001; Ostrom, 1990)
Public interest oversight and prevention of waste	1919s through 1950s: • Conservation laws and public oversight of well spacing (Olien & Olien, 2002) • Regulations limit extraction to the "maximum efficient rate" to prevent reservoir depressurization (Weaver, 1986) • Regulations prevented the waste of gas through flaring (Weaver, 1986) • Well-construction rules to prevent cross-contamination (Weaver, 1986)	1900s through 1980s: • States adopted statutes codifying common law principles • Concepts of "safe yield," "sustainable yield," and "perennial yield" are used to set a production limit from aquifers (Bredehoeft, 1997) • Well-construction standards introduced to protect public health and prevent contamination
Coordinated management of the resources as a whole and mitigating incidental impacts to some users	1930s through Today: • Scientific understanding of subsurface fluid mechanics and optimization • Recognition of a broader system of interconnected components within an oil and gas reservoir • Technological advancement in injection wells allow manipulation of fluid flows and minimization of waste (Weaver, 1986)	1980s through today • Groundwater modeling advances • Increasing legal recognition of surface water and groundwater connections • Increasing utilization aquifer recharge and artificial storage and recovery technology • Efforts to foster collaborative and regional groundwater planning, like California's

Table 1.5. (*cont.*)

Issue	Oil and gas	Groundwater
	• Realization that state and national laws could not compel landowners to cooperate effectively (Hardwicke, 1948) • Unitization contracts equitably distribute benefits of the common reservoir (Weaver, 1986) • Technologies like horizontal well drilling and fracking render the theory of "peak oil" obsolete (Cavallo, 2004; Jarvis, 2011)	Sustainable Groundwater Management Act • Claims that society may have reached or surpassed "peak water" (Gleick & Palaniappan, 2010; Jarvis, 2011)

progressive era before World War II, legislatures and agencies attempted to protect the substantive public interest and private self-interest by regulating and placing limits on resource extraction (Horwitz, 1982). Finally, the patchwork of rights, laws, and regulations designed to solve these resource development issues failed to optimally use the resource as a whole, especially as development began to interfere with competing interests and interconnected resources.

When comparing the major periods of the development of these two resources, similar approaches and similar issues arose. During each period, the issues and legal solutions appear quite similar until one reaches the current period, reflected in Table 1.5. Today, at least in the oil and gas context, the protection of "public interest" has to some extent become "redefined as simply a reflection of the sum of the vectors of private conflict" and an "outcome of competition between interest groups" as expressed in unitization agreements (Horwitz, 1982). Unitization agreements formed as the default management tool, due to the perception that state and federal regulations were unlikely to foster efficient and cooperative development of oil and gas reservoirs (Hardwicke, 1948).

However, absent from the groundwater column in Table 1.5 is something akin to unitization agreements and a recognition that laws and regulations can be supplemented with private contracts and aquifer-wide agreements.

Later in this book, we discuss several examples that could be considered "near misses" for aquifer unitization, showing a potential trend that governance of aquifers and groundwater would, again, follow the oil and gas industry's example.

We also follow the tracks laid down by the oil and gas industry and imagine the shape, dynamics, and issues of applying the unitization model to aquifers. We investigate the terms of unitization agreements, elaborate on key conceptual elements (like determination and redetermination) that could be contributed to current aquifer governance theories, and suggest that aquifer unitization agreements (or something like them) may emerge as the standard governance approach, just as they have for the domestic and international oil and gas industry.

2

Overview of the History of Collective Action on Subsurface Resources

Nature made these pools units, and it was therefore impossible to operate them economically and efficiently except as units
—*H. L. Doherty, 1939, in* New York Times *(1939)*

Conservation measures should have been adopted long ago, for the ultimate result has been obvious for many years, but it has been only recently that the rate of depletion has increased so rapidly as to become a matter of acute alarm as to whether conservation measures can be adopted quickly enough to avoid embarrassment.

The above quote could have easily been attributed to Cuthbert et al. (2019) as part of the environmental "time bomb" they predicted for the replenishment of global groundwater systems and climate change. Yet the quote is from Doherty's (1924) letter to President Coolidge (quoted in Hardwicke, 1948, p. 181). At age fifty-four, Doherty was sounding the earliest alarm about "peak oil" before the many other "peaks" predicted to come, in terms of subsurface resources meeting the needs of humans and nature (Jarvis, 2011).

Given that by 2019 the United States was one of the world's leaders in oil production, it was surprising that Doherty was calling for slowing the rate of oil production nearly a century earlier. As outlined in Chapter 1, his concerns were related to national security, later summarized by others as "wasteful haste"; current parallel concerns for groundwater and aquifers are considered "undesirable results." Some might say that historic concerns about oil and contemporaneous concerns about groundwater and aquifers may defy the idiom that oil and water don't mix.

As practicing groundwater professionals, we observed the mechanical connection between pumped groundwater and subsurface storage firsthand, as described in Jarvis (2011). What surprised us was the apparent lack of policy on conserving the "common pool" – the groundwater *and* the aquifer storage. We discussed the challenges that were emerging from the apparent lack of conserving the aquifer "unit" to groundwater development in the early 2000s with former land managers in the oil and gas industry, who expressed surprise that groundwater was not "unitized" to conserve the aquifer in the same way unitization was designed to conserve the oil and gas "reservoir."

At the time, we had never heard of unitizing groundwater or aquifers – so began our research in 2003. The outcome of the early research was pitching the idea at international conferences on nonrenewable groundwater and groundwater governance as the Draft Articles of the Law of Transboundary Aquifers were introduced by the International Law Commission in 2008. Revisiting Article 2 outlining the use of terms indicated that *"utilization of transboundary aquifers or aquifer system includes extraction of water, heat and minerals, and storage and disposal of any substance"* (emphasis added). We saw immediate application of the notion of unitization as initially proposed under Doherty's plan because unitization was already used to manage geothermal resources and sequester carbon in the subsurface (Jarvis, 2011).

While many legal, economic, and political science scholars have dedicated their careers to the theory and practice of governing the commons, we have not encountered biographical summaries of the pioneers in collective action of the subsurface. The following sections introduce two of the leaders in this area: Henry Latham Doherty, an industrialist who championed collective action in the subsurface as part of private enterprise; and Elinor Ostrom, the Nobel Prize Laureate in Economics well known for her work associated with the book *Governing the Commons*, but less so for her PhD dissertation focusing on collective action of the subsurface as part of public entrepreneurship.

2.1 Who Was Henry L. Doherty?

We have introduced the notion of unitization and the principal proponent of the concept, Henry L. Doherty, but who was this person and why should the reader care? Beyond snippets of information in historic newspapers, his obituaries, and online encyclopedias maintained by historical societies, we found little in the way of biographical information about Henry Latham Doherty (1870–1939).[1]

Doherty's story is typical of the "pull yourself up by your own bootstraps" idiom. He left school at the age of ten, sold newspapers, and worked as an office boy for a gas company in Ohio. According to Campbell (1925), Doherty rose to company management by age twenty-one and was chief engineer at thirty, a

---

[1] Doherty is not to be mistaken for Harry M. Daugherty, who was the Attorney General of the United States under Presidents Warren G. Harding and Calvin Coolidge, and was ultimately disgraced for his involvement in the Teapot Dome scandal during Harding's presidency.

millionaire at thirty-five, and a billionaire by current valuations at age fifty-five. He died at the age of sixty-nine as a lifelong sufferer of rheumatoid arthritis.

Doherty founded the Cities Service Company in 1910, an oil and gas company with two hundred subsidiaries nationwide, which eventually became today's CITGO Petroleum Corporation. His real estate holdings in Florida included a chain of hotels, and he was credited with reviving the tourist business in Florida after the Great Depression. He also owned land and erected buildings in the financial district of New York.

Doherty was a founding member of the American Petroleum Institute, a national trade association active today and whose mission included funding and conducting research, establishing industry standards, as well as influencing public policy to support the US oil and gas industry. By 1930, he had more than 150 patents. According to Doherty's obituary published in the *New York Times* (December 27, 1939), he "derived the greatest satisfaction from the results of his five-year fight for a program of oil conservation, known as 'unit operation' of each new pool."

What inspired Doherty to promote the notion of unitization? Our research does not point to a documented philosophy. However, Doherty was about forty years old when Gifford Pinchot's *The Fight for Conservation* was first published in 1910. Doherty may have been inspired by Pinchot, who wrote:

Many oil and gas fields, as in Pennsylvania, West Virginia, and the Mississippi Valley, have already failed, yet vast amounts of gas continue to be poured into the air and great quantities of oil into the streams. Cases are known in which great volumes of oil were systematically burned in order to get rid of it.

(pp. 7–8)

In his 1924 letter to President Coolidge, Doherty references "conservation" with some parallels to Pinchot's definition of conservation as including development (Hardwicke, 1948, p. 181), but not necessarily Pinchot's "husbanding of resources for future generations" (Pinchot, 1910, p. 42). Conservation of resources in the eyes of Pinchot was extended to the prevention of waste. He also promoted the idea that natural resources must be developed and preserved for the benefit of the many, not merely for the profit of the few; conservation meant the greatest good to the greatest number for the longest time. Pinchot also wrote that "conservation advocates the use of foresight, prudence, thrift, and intelligence in dealing with public matters for the same reasons and in the same way we each use foresight, prudence, thrift and intelligence in dealing with our own private affairs" (Pinchot, 1910, p. 48).

Doherty's pitch to President Coolidge was that conservation must focus on ownership to be determined by other means than simply "capture," that the laws for oil and gas conform to laws governing other property, that the division among different landowners be made with a greater degree of equity, that the development of pools without haste could recover double that under the rule of capture, and that more oil could be raised to the surface without the cost of pumping by merely preventing unnecessary waste of the gas drive in the oil reservoir, instead increasing the recovery rate from 10 percent to 35 percent to nearly 100 percent. He envisioned a model where the separate states passed legislation providing for "oil districts," similar to the laws that have been passed for irrigation and drainage districts. He recognized that his proposed changes would be difficult to implement because "every business and industry is controlled largely by its conservative and standpat element. Governmental changes largely come from the radical and irresponsible element. Every industry fears to invite government interference for fear of the radical and irresponsible element and impractical reformer" (Doherty Letter to President Calvin Coolidge dated August 11, 1924, in Hardwicke, 1948, p. 185).

Doherty recognized the challenge that lay ahead, with "fear it will have to be done by our Federal Government without much help from the men in the oil industry, and with the determined opposition of some of the men in the oil industry." (Doherty Letter to President Calvin Coolidge dated August 11, 1924, in Hardwicke, 1948, p. 189.) Doherty thought the depletion issue was so important that he recommended it receive attention from a cabinet officer. He also thought that the Bureau of Mines and the United States Geological Survey were important (due to the scientific and technical nature of depletion issue). Thus was the beginning of Doherty's plan.

In 1927, the American Petroleum Institute adopted unitization; however, the concept was not universally accepted by leaders in the oil industry who were primarily concerned with Doherty's plan that the petroleum ownership laws be changed so that "ownership of oil would vest in common in all of the land situated over a pool and that no owner should have the right to drill independently" (*New York Times* 1939). Doherty argued that orderly development of oil pools would lead to maximum recovery: "Nature made these pools units, and it was therefore impossible to operate them economically and efficiently except as units" (*New York Times*, 1939). There were scathing rebuttals by prominent oil and gas geologists L. C. Snider and E. DeGolyer (Lovejoy, E. DeGolyer and Sinsheimer in Hardwicke, 1948). Doherty apparently had little patience for antagonists against the notion of a federal oil regulatory law who based their arguments on the constitutionality, or takings, of landowner rights. He argued that no one could use their property in a way that was injurious to the property of a neighbor (*New York Times*, 1939).

DeGolyer later became a proponent of unitization and served on the H. L. Doherty Memorial Fund Committee, which led to the publication of the Henry

L. Doherty Memorial Volume of *Petroleum Conservation* by the American Institute of Mining and Metallurgical Engineers in 1951 (Buckley, 1951). DeGolyer noted that the "expansion of the dissolved gas is in many fields the single important agent and in most fields a potent agent which moves the solution to the well," independently" (*New York Times* 1939). Doherty argued that orderly development of oil unit and should be operated as such regardless of divisions of ownership, he [Doherty] became the great protagonist of unit operation" (Lovejoy, DeGolyer and Sinsheimer quoted in Hardwicke, 1948, p. 257).

In an Oklahoma Historical Society Encyclopedia online posting, Weaver (undated) writes that "in keeping with his scientific approach to business Doherty, who preferred to be called chief engineer rather than president, established the Doherty Research Company (DORECO) at Bartlesville in 1916 to advance his theories of field unitization as a means of oil conservation and as a location to train petroleum geologists and engineers." Similarly, the Lamont–Doherty Earth Observatory at Columbia University was named after a major contribution from the Henry L. and Grace Doherty Charitable Foundation, a major supporter of institutions studying another common pool resource, the oceans.

2.2 Who Was Elinor Ostrom?

Unlike Henry Doherty, Elinor Ostrom's background is well documented, given the timing of her contributions and publications. Many tributes were published following Ostrom's death in June 2012. One of our favorite tributes was by environmental writer James Workman, who referred to Elinor Ostrom as "The Patron Saint of Enviropreneurs" (Workman, 2012).

Elinor Ostrom was raised in southern California. After graduating with bachelor's and doctoral degrees from UCLA in the mid-1960s, Ostrom lived in Bloomington, Indiana, and served on the faculty of Indiana University. She earned a Nobel Prize in 2012 for her work on common pool resource management and institutions for collective action. A Workshop on the Ostrom Workshop (WOW) continues to be sponsored at Indiana University to celebrate both her and her work. Workman succinctly summarized her work: "Economists obsess about incentives; Ostrom, like enviropreneurs, orients us toward institutions" (Workman, 2012). Fennell (2011) coined "Ostrom's law": "A resource arrangement that works in practice can work in theory" (p. 10). We felt this "law" captured the nuance associated with governing the subsurface, regardless of whether the enterprise setting has a private or public entrepreneurial focus.

While most readers are familiar with the body of literature by Ostrom on the commons and related institutions from one of her most widely cited works,

Governing the Commons (Ostrom, 1990), what many readers may not know is that groundwater was the common pool resource profiled in her doctoral dissertation (1965). In that remarkable and underappreciated tome, Ostrom underscores the importance of transdisciplinarity in groundwater and aquifer management, as well as lays the foundation for the future of water governance in her backyard of southern California and elsewhere in the western United States. The notion of water conservation during the 1960s focused on the dictum "the control of nature is won, not given," the inscription over the entrance to the University of Wyoming Engineering Building that served as the inspiration for the title of a John McPhee book. The message is the water management paradigm of "Total Water Management" – a drop of water that makes it to the ocean is wasted water. Water conservation in the early days of European settlement of the western United States meant building dams, levees, and channels to control floods, generate power, and develop previously unusable wetlands. It also focused on mega water development projects such as the North American Water and Power Alliance pitched by the Parson Corporation based in Pasadena, California, and storing water in dams as depicted on postage stamps during this time period. Saltwater intrusion in the West Coastal groundwater basin, upon which the Los Angeles metropolitan area overlies, began the journey of Ostrom's "enviropreneurialism" to seek public solutions to a "commons" problem.

At the time of Ostrom's dissertation research in the 1960s, the West Basin water industry had 77 enterprises out of the previous 310 enterprises inventoried in 1945. Oil companies constituted a group of industries using water for their own use – primarily oil refining. The proportionate share of the total yield of the basin was 36 percent, yet the oil companies were only able to meet approximately 50 percent of their freshwater needs using their own production; the rest was purchased from neighboring water entities. According to Ostrom (1965, p. 51), "the oil companies have been very concerned about the preservation of the basin as source of high quality water for specialized industrial processes."

Other industrial water users included a paper company, steel manufacturing, and chemical manufacturing, among others. Their production approached 5 percent of the total production from the West Basin. Like the oil companies, the industrial producers also purchased water from public and private water agencies.

Nonindustrial users, such as Los Angeles County, a turf club, real estate concerns, schools, and farmers produced about 5 percent of the total production from the basin. Purchases from public and private water agencies were common.

Municipal water departments, related districts, and county waterworks supplied water to individuals and firms within their boundaries and produced about 21 percent of the total production from the basin in the 1960s.

Six private water companies were active water producers in the West Basin during the 1960s. And while the demand for water increased three-fold over a period of approximately twenty years, the bulk of the water supplied by the water companies was derived from imported water from wholesalers such as the Metropolitan Water District, and reclaimed water.

The Los Angeles County Flood Control District was charged with flood control and conservation of flood and waste waters in the 1960s, and as such, was an active player in the management of the West Basin through replenishment. The Replenishment District was eventually created to purchase water injected or spread in the basin for recharge for seawater barrier construction and control.

The California Water Resources Department served as the Watermaster to enforce a 1961 judgment and agreement in the West Basin. According to Ostrom (1965), the Watermaster-insured water producers withdrew an agreed amount of water each year, as well as administered an "exchange pool" that permits water producers to purchase rights to additional pumped groundwater from others who gave up equivalent rights and substitute imported water for groundwater. At the time, the Watermaster had the resources to check pumping equipment and records, check water levels in wells, and assess the status of saltwater intrusion. This data was summarized in a report that was available to all agencies who had regulatory actions in the basin.

Replenishment districts were created to (1) establish a balance between the supply and demand for groundwater, (2) eliminate the threat of saltwater by financing injection of freshwater into barriers, and (3) replace water that had been pumped in excess of natural replenishment. The replenishment districts apparently desired cooperative action, but could undertake regulatory controls on their own without seeking joint solutions.

The role of water associations was important by serving as the forum for concerns on gaining control of saltwater intrusion and developments toward groundwater management of the West Basin. The basin association developed a formal structure for "sustained negotiation and communication by all affected parties" (Ostrom, 1965, p. 80). The water association also served a role by representing local interests before political bodies.

By the time Ostrom's research was almost completed in the mid-1960s, there was both a recognition that other nearby groundwater basins were hydraulically connected to the West Basin, thus not only requiring public enterprises to manage the interconnected basins, but also consolidation of water companies into fewer, larger companies, as well as an elimination of many private water providers (Ostrom, 1990). The consolidation created "the incremental development of a superstructure composed of diverse public enterprises, private associations and private engineering and law firms" (Ostrom, 1965, p. 86).

Ostrom recognized the efforts to conserve the West Basin aquifer; she notes that the West Basin producers

adopted a strategy which involved searching for a new organizational arrangements or institutional devices that would enable them to solve most of their problems at the local level and provide for an expanding level of water services to the entire community. *This strategy resembles that of a private entrepreneur who creates new organizations in order to join together diverse products and operations in developing a more effective enterprise.*
(Ostrom, 1965, p. 87, emphasis added)

Ostrom concluded that "the strategy followed by individuals in West Basin might be viewed as a series of undertakings in public entrepreneurship" (Ostrom, 1965, p. 88).

In many respects, the process, cooperation, and resulting superordinate identity created by the desire to conserve the southern California groundwater basins by the public enterprises closely coincide with the unitization process proposed by Doherty for the private enterprises in oil sector. Both enterprises relied upon decentralized institutions for "conserving" the reservoirs and aquifers within designated geographic, and geologic, boundaries. Both enterprises relied upon experts to guide extraction and redetermine shares of the resources. And, both enterprises' professional associations share knowledge and influence public policy to support their enterprises.

More importantly, Ostrom's dissertation and follow-up work by her protégés Blomquist (1992) and Schlager (2004), among many others, would set the stage for one of the most ambitious experiments managing the groundwater commons – the California Sustainable Groundwater Management Act of 2014.

2.3 The 5P Framework

While much has been written about groundwater governance at the global scale, as summarized by Villholth, Lopez-Gunn, Conti, Garrido, and van der Gun (2018), what was missing from the models of groundwater governance was an emphasis on the "aquifer": the groundwater storage container. As part of the treatise on *Advances in Groundwater Governance*, van der Gun and Custodio (2017) inventoried subsurface human activities, recognized the fragmented governance, and offered suggestions for governance. However, they imply that the commercial goals of the *private* enterprises such as the oil, gas, and mining sectors of pursuing profits are somehow at odds with the pursuit of wealth created for society through *public* enterprises, suggesting the two enterprises cannot peacefully coexist.

In 2011, we proposed the four core principles – or "4P" framework – behind the unitization of aquifers (see Jarvis, 2011), which have since been acknowledged by entrepreneurs in both private and public enterprises and have been the subject of modern lawsuits:

(1) *Promote* groundwater exploration and development in underutilized areas, for example, in "megawatersheds" that are being promoted as a new exploration paradigm. For example, Martin-Nagle (2016) suggested unitization as the best legal regime for the development of offshore aquifers.

(2) *Preserve* the storage capacity of aquifers by promoting local control of groundwater development. For example, the Agua Caliente Band of the Cahuilla Indians ("the Tribe") in California has seen the potential need to establish a right to the pore spaces, not the water alone, in the aquifer below their reservation. The Tribe filed a lawsuit on May 14, 2013, against Coachella Valley Water District and Desert Water Agency, as will be discussed later in the book.

(3) *Private* investment in the "post-modern hydrologic balance" including aquifer storage and recovery (ASR); managed recharge (similar to secondary and tertiary recovery operations used in the oil and gas industry); nonrenewable groundwater, which does not fit well within the paradigm of integrated water resources management, as well as other opportunities such as remediating contaminated groundwater; ecosystem services; and the spirituality of water. For example, Tuthill and Carlson (2018) describe marketable Aquifer Recharge Units (ARUs) in the Eastern Snake Plain Aquifer in Idaho.

(4) *Prevent* disputes by "blurring the boundaries," thus creating a new community of users with a superordinate identity who agree on how to "share" groundwater and aquifers. For example, Blumberg and Collins (2016) describe the transformation of a Texas groundwater conservation district through a water-management solution based on available water volume beneath properties.

In Chapter 1, we introduced a fifth core principle that has emerged following some of our international work on transboundary aquifers: (5) *Principled* collaboration, which encourages good faith commitments among legal, technical, and social working teams.

2.4 Application of Unitization and Collective-Action Principles through Time

We inventoried a broad spectrum of expertise that relied on the principles of unitization and collective action when governing subsurface resources as listed in Table 2.1. Business-linked private enterprises, political scientists linked to private and public enterprises, economists, legal scholars and lawyers, geographers, geologists, and engineers have either directly mentioned unitization principles, or indirectly linked to unitization and collective-action concepts,

Table 2.1. *Inventory of authors applying unitization principles*

Expertise	Emphasis	Year	Citation
Business	Private entrepreneurship Oil and gas	1920s–1940s	Doherty
Political Science	Public entrepreneurship Groundwater	1965	Ostrom
Economics and Law	Geothermal resource management	1977	Sato and Crocker
Public Policy	Groundwater governance	1992	Blomquist
Economics	Groundwater	1997	Anderson and Snyder
Economics	Groundwater Oil and gas	2005–2008	Libecap
Economics	Aquifer communities	2009	Shah
Law	Carbon sequestration	2009	Flatt
Law	Groundwater management	2011	Clyde
Geography	Groundwater governance	2011	Linton and Brooks
Geology	Aquifer governance	2011	Jarvis
Law	Water and human rights	2011	Carlson
Law	Offshore aquifers governance	2016–2020	Martin-Nagle
Public Policy	Groundwater (3-D management)	2016	Blumberg and Collins
Law and Water Policy	Collective action for aquifer space	2018	Wiley
Engineering and Business	Aquifer recharge units	2018	Tuthill and Carlson

to subsurface resource management and governance. Van Laerhoven and Berge (2011) found a comparable spectrum of expertise that relied on Ostrom's *Governing the Commons* (1990) to develop their arguments on problem-solving within public enterprises, including economists, sociologists, anthropologists, political scientists, legal scholars, geographers, biologists, ecologists, foresters, hydrologists, and students of public administration. It is clear subsurface contains multiple resources that are bound within an interconnected system. We refer to these as *transresources* that are inexorably linked with a focal natural resource and must be addressed within any collective action or aquifer governance approach through a lens of *transdisciplinarity*.

3

Governance of Groundwater and Aquifers

You Can't Separate One from the Other

> Good governance of groundwater is still lacking. There is a huge gap in
> public and private institutions dedicated to groundwater, which does not
> allow a proper governance of this valuable resource.
>
> —*A. Rivera (2019)*

The current era of groundwater governance shows that the top-down management
model is not the only effective means of addressing natural resource issues.
Governance approaches have taken many forms, including international treaties,
legislation, novel uses of traditional water districts, and intergovernmental agreements. These approaches to governance, however, have sought only to address
groundwater availability, not aquifer potential. The limited perspective of these
approaches has also limited the flexibility, issues included, and scale of the problem.
As introduced in Chapter 2, we consider the most comprehensive approach to
subsurface resource governance to be found in the context of unitization.

Expanding the governance scope from groundwater to aquifers also raises new
issues, complicating the theoretical and practical implementation of any management system. While groundwater is often treated as a publicly owned resource, the
aquifer storage space is presumably a private property right. An aquifer is an
intersection of public and private uses, which will challenge any groundwater
governance approach in the future. Likewise, aquifers contain multiple
transresources that are bound within an interconnected system and must be
addressed within any aquifer governance approach. This includes groundwater
transresources, which must be included in the governance system to holistically
address the aquifer system.

This chapter will view aquifers and their governance through the lens of systems
thinking, exploring the concepts as interconnected web of causes, effects,
feedbacks, and emergent properties (Daniels & Walker, 2001). Systems thinking
is based on nonlinear approaches to complex issues. Systems are made from

elements, relationships, boundaries, environments, inputs, and outputs (Daniels & Walker, 2001). Systems can transform into new systems, create feedback loops, and produce emergent properties as results of the system (Daniels & Walker, 2001). Positive feedback loops strengthen the feedback, while negative feedbacks restrain the dynamic between two elements. Systems thinking allows for transdisciplinary and transresource effects to be explored, breaking through the analytical challenges presented by linear approaches.

Two general systems will primarily make up the systems analysis: the social-legal system and the resource system. These two systems create two domains, akin to a system boundary, that must both be embraced within an effective governance system. The interaction between social and environmental systems has been discussed by Linton and Budds (2014), creating the hydro-social system, which proposes that water and society form a dialectic relationship. A similar system approach was developed by Ostrom, Cox, and Schlager (2014) called the social ecological system (SES), which links human and resource governance in an analytical framework. Rights, rules, and organizations form components of the SES framework, which become variables in a predictive analysis. Santelmann et al. (2012) described watersheds as complex adaptive systems using a social-ecological system framework. This chapter utilizes the concept of interrelated social and resource systems as the foundational concept underlying governance of aquifers and their transresources.

First, we will describe the emerging issues that require a transition from groundwater to aquifer governance. This chapter ends with a discussion of various examples of current groundwater governance approaches, while also viewing the examples from an aquifer governance perspective. The successes of these examples will be shown, but we will also use the aquifer governance perspective to expose the aspects that are missed when other groundwater governance approaches are used. The chapter also shows that unitization provides a middle path between the public–private debate and forms a viable example of a holistic governance approach that encompasses both the social and resource scales.

3.1 The Need to Transition from Groundwater Governance to Aquifer Governance

The transition from groundwater management to groundwater governance expanded the scope of the legal, social, and participatory aspects of groundwater. Likewise, the scope of the included resources governed needs to be expanded to the various aspects of an aquifer. The definition of an aquifer is not settled in governance and is often conflated with groundwater.

The definitions of an aquifer are manifold and, according to the classic text on groundwater by Freeze and Cherry (1979),

it means different things to different people, and perhaps different things to the same person at different times. It is used to refer to individual geologic layers, to complete geologic formations, and even to groups of geologic formations . . . . An aquifer is best defined as a saturated permeable geologic unit that can transmit significant quantities of water under ordinary hydraulic gradients.

(Freeze & Cherry, 1979, p. 47)

Because aquifers have the *potential* to store water, it is sometimes assumed that aquifers are *merely* for the storage of water, and nothing else. For example, the Draft Law of Transboundary Aquifers defines an aquifer as both the "water bearing formation" and the groundwater within that formation, including both the territorial (pore spaces) and the transient portions (groundwater) (International Law Commission, 2008; McIntyre, 2011). Even more complex, an "aquifer system" contains multiple discrete aquifers and other related resources (International Law Commission, 2008). The commentary to the DLoTA includes multiple definitions, some referencing strata of water-bearing zones, others referencing porosity and permeability (International Law Commission, 2008). However, as introduced earlier in this book, the DLoTA also mentions heat extraction, carbon storage, waste disposal, and minerals as components of aquifer systems (International Law Commission, 2008; Jarvis, 2011). Clearly, aquifers are much more than groundwater and are part of a complex system of heat, minerals, storage spaces, and many other transresources. The governance of groundwater alone cannot accomplish the effective governance of all transresources.

3.2 The Transresource Problem: Groundwater versus Aquifer Governance

Aquifers contain groundwater, and so much more. The traditional and groundwater governance paradigms treat most issues related to aquifers as groundwater issues. Under the traditional paradigm, issues of groundwater contamination, stream interactions, availability, subsidence, and seawater intrusion are indirectly managed through the narrow applications of conditions and restrictions on groundwater rights. The groundwater governance paradigm incorporates participatory, inclusive, and collaborative methods to reduce conflict and improve outcomes (Villholth & Conti, 2017). Both paradigms result in regulations on groundwater, only indirectly addressing the transresource issues prompting a policy intervention.

As discussed, transresources are physically interconnected resources within an aquifer system. For example, aquifers contain groundwater, storage spaces, heat, contamination by natural- and human-introduced contaminants, biological components, hydrocarbons and gases, minerals, and chemical processes; each resource is a transresource. For aquifers, groundwater is only one transresource among many, but it is often the only consideration under traditional and groundwater governance paradigms. Aquifers are interconnected with surface water through recharge, capture, and baseflow. The previous paradigms often attempt to address issues with these transresources with one approach: reduce consumption of groundwater until the issue is resolved. These paradigms restrict, rather than unlock, the potential benefits of other solutions that direct address transresource issues.

Aquifer governance attempts to incorporate the entire aquifer system into the governance system. As seen in Figure 3.1, aquifers contain a variety of resources and elements that form a dynamic system, and groundwater forms only a small but important section of the whole. Under the traditional and groundwater governance view, groundwater plays the central role in the physical system, with other transresources playing only a secondary role to the other aspects of the aquifer. Groundwater and surface water create a subsystem within the overall system; allowing water to flow between them is recognized as conjunctive management in groundwater regulations (Cobourn, 2011; Getches, Zellmer, & Amos, 2015). Issues of contamination in an aquifer by biological and other elements are linked to the aquifer system by potential in situ (in place) bioremediation (United States Environmental Protection Agency, 2013). Pore spaces and aquifer storage provide the medium for not only groundwater and related managed recharge, but also for carbon dioxide and waste storage; however, they may also play a critical role in subsidence. Aquifers have cooling and heating properties and could potentially be

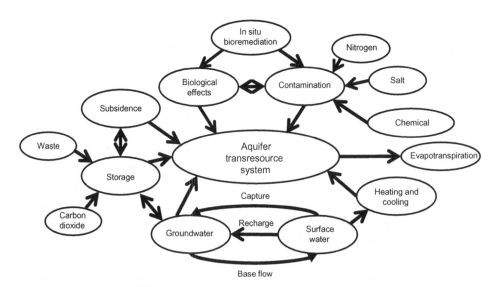

Figure 3.1 The aquifer transresource system

utilized as heat sinks for cooling buildings (Birks, et al., 2013). Discharge of groundwater into rivers as baseflow plays a significant role in maintaining cooler river temperatures for salmonid habitat (Essaid & Caldwell, 2017). The primary water loss from the aquifer transresource system (Figure 3.1) comes from evapotranspiration from plants, water bodies, soils, and discharge to oceans. For the sake of simplicity, we consider the ocean to be a form of brackish surface water, as its role in the aquifer transresource system is similar. Ostrom (1965) similarly made no fundamental distinctions between rivers, lakes, and the ocean, other than their salinity. These effects and influences can be quantified and incorporated alongside groundwater within the governance system, but have only been indirectly addressed through groundwater governance.

Likewise, groundwater governance approaches have missed opportunities for creating a more holistic governance system by failing to address the bottlenecks precluding the effective use of regulation to change water users' behavior. As groundwater management has moved to groundwater governance, collaborative, participatory, and additional levels of jurisdiction have been incorporated into the traditional system of regulations and laws (Villholth & Conti, 2017). Social and legal constraints on access to groundwater form the primary method of preventing harm by water users, as shown in Figure 3.1. These constraints come from a variety of sources, including critical area designations, wellhead protection zones, sole source aquifer designations, and other restrictions placed on land overlying aquifers and groundwater use. The conditions placed on water rights are another source of groundwater regulation, limiting use to a certain volume and season, in some cases. Water users' activities are primarily determined through forms of groundwater rights under one of five different doctrines (Getches et al., 2015; Mondo, 2018).

However, the status of water rights as a property right has threatened to undermine social-legal constraints, amounting to a public taking of private property requiring compensation (Bonini, 2018; Griggs, 2014; Newman, 2012). This argument has hampered the effectiveness of social-legal constraints, undermining their effect on water users. Additionally, restrictions on access to groundwater could implicate human rights. The emergent effects from this system have been conflict between water users and agencies, ineffective attempts to control overuse of groundwater, and a disconnect between the resources and the physical availability of water, as seen by the lack of connection between the water availability subsystem and groundwater rights (Figure 3.2).

Groundwater governance has introduced negotiation and collaborative govern-ance into the system, giving water users access to the process of the development and application of social-legal constraints to their groundwater rights (Villholth & Conti, 2017). Adaptive management has been suggested for use in the planning

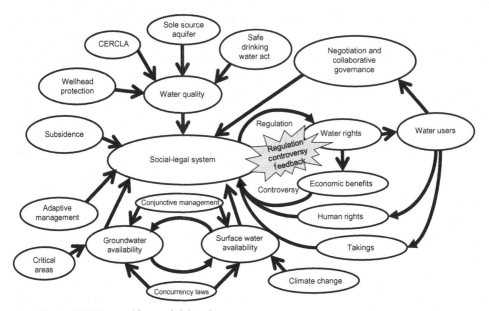

Figure 3.2 The aquifer social-legal system

and administration of water rights (Chaffin, Craig, & Gosnell, 2014; Cosens et al., 2017): "Adaptive management is a useful tool in a problem-solving approach and incorporates a step involving reflection and learning" (Cosens et al., 2017, p. 8). Adaptive management works best in situations where science-based adjustments are made in uncertain and controllable situations (Cosens et al., 2017). The approach relies on "iterative cycles of goal determination, model building, performance-standard setting, outcome monitoring, and standard recalibration" (Craig & Ruhl, 2014, p. 8). Adaptive management creates a feedback loop for planning and the incorporation of new information into future decisions. However, adaptive management, along with other governance concepts, becomes mired in the "regulation-controversy feedback" during implementation. The dampening effect of the Takings Clause of state and federal constitutions, along with human rights, creates strong countervailing forces against regulation, even when collaboratively developed and adaptively managed.

The effects of the regulation-controversy feedback hinder the effectiveness of social-legal constraints, making the feedback a bottleneck for the whole system. Additionally, groundwater rights are only indirectly associated with groundwater availability. Further, subsidence and other water quality concerns may only be addressed through regulation of groundwater rights within the current system. Subsidence can result in a loss of aquifer storage space (Jarvis, 2011). The loss of storage spaces is not often seen as a property rights issue, but could potentially emerge as one in the coming years.

These disconnections between the negative externalities and the rights to groundwater are substantial factors creating the emergent properties of the system. The concept of transresources bridges the disconnections found the in the social-legal system of groundwater governance, creating aquifer governance. These disconnections form the second major reason for a needed transition from groundwater governance to aquifer governance discussed in the next section of this chapter: the characterization of groundwater as either public or private.

3.3 The Public–Private Aquifer Paradox: Private–Public Myths

Aquifers create a public–private property paradox, but may actually be neither public nor private property. Authors often claim that water (and groundwater) is publicly owned (Getches et al., 2015). Some authors claim that the implementation of groundwater banking has privatized otherwise public rights in aquifers (Keats & Tu, 2016). Other authors claim that the ownership of the overlying land creates a possessory interest in the groundwater below the landowners' estate (Newman, 2012). None of these arguments address the right to possess storage (or pore) spaces within the aquifer, whether currently or potentially filled. Some states have statutorily recognized pore spaces as private property (Montana, Wyoming, North Dakota), while other parties have argued that aquifer pore spaces are public resources, available for use by all (*Agua Caliente Band of Cahuilla Indians* v. *Coachella Valley Water District* et al., 2018). Therefore, the container is often private, but its use and contents are public. Somewhere between all these positions lie the transresource system and various rights to the different transresources.

Despite the popularity of the public–private debate, property comes in four forms: common, private, public, and collective. Common property is held by all, and no individual system of rights has been established (Barbanell, 2001; Peredo, Haugh, & McLean, 2017). Open access to the resource often produces the tragedy of the commons, where individuals benefit from the resource, but do not feel the negative effects of overuse (Libecap, 2008; Ostrom, 1990). No market-based mechanism discourages collective overuse of the resource (Barbanell, 2001; Hardin, 1968). Often, privatization of the resource is suggested as a solution to the tragedy of the commons (Libecap, 2007, 2008). Private property rights are molded to link negative externalities to the social harm caused by their use (Libecap, 2008). However, designing property rights is difficult, given that predicting the negative and positive externalities requires a high degree of knowledge of a resource (Barbanell, 2001). Public property, however, gives possession to a state agency and the "asset or resource is owned by a governmental authority" (Peredo et al., 2017). Public property is accessible by all, like common property, but a state holds the property. Collective property

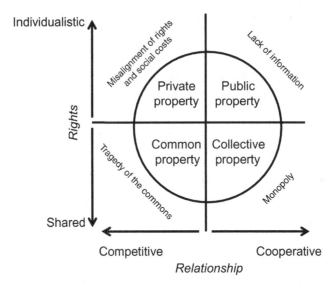

Figure 3.3 The property regime arc: Property regimes lie on a spectrum of rights and relationships

combines aspects of the other forms, recognizing individual shares of a commonly held resource (Peredo et al., 2017). Under a collective property regime, rights are shared as a group and governance is cooperative rather than competitive, as under common and private property (Foster, 2011).

As shown in Figure 3.3, each of these forms of property has an associated policy issue that considers negative externalities. Common property regimes produce the tragedy of the commons, since open access defines individual rights, but each actor competes over the resource. Private property regimes do not always internalize social costs for their use. Public property regimes place ownership in the state, which can suffer from a lack of information about how to best manage the resource to achieve social benefits. Collective property uses shared rights and cooperation to manage a resource, but cooperation can also lead to monopolies, unreasonably restrictive burdens on commerce, and socially unfair advantages (Scott, 2016).

Some economists assert that overuse of a resource is considered a result of a failure to internalize the negative externalities of consumption (Singh, 2016). Two solutions may be used to address this mismatch of benefits and costs of resource use: the Pigouvian Tax-Subsidy Approach and Coasian Property Rights Approach (Singh, 2016). A Pigouvian Tax places an additional cost to extract groundwater, which is intended to reduce its consumption (Schuerhoff, Weikard, & Zetland, 2013). The latter can be accomplished by the assignment or reassignment of property rights that internalize the externalities of resource consumption (Lai, Davies, & Lorne, 2016). Several authors have suggested privatizing groundwater as a means to accomplish the restoration of groundwater sustainability (Anderson, 1983; Ghosh & Willett, 2012; Provencher, 1993; Provencher & Burt, 1994; Smith, 1986). Each of these proposes converting current groundwater rights into in situ private rights, each raising issues related to management of the aquifer overall. These concepts ignore that water rights themselves meet the definition of a property right: "a bundle of entitlements defining the resource owner's rights, privileges, and limitations for use of the resource" (Ghosh & Willett, 2012, p. 144). At least in a practical, economic, nonlegal sense, water rights constitute a property right insomuch as they grant conditional control of a resource to individuals.

The issue is not whether water rights constitute property rights in a theoretical and legal sense, but whether water rights adequately internalize externalities and need reformation. Property rights can be changed through edict (potentially creating constitutional takings issues) or by contract, called "Coasian Bargaining" (Lai et al., 2016). The creation or modification of property rights has two meanings: "It can mean either the establishment of private property rights over previously common property, in which there is a phenomenon of rent dissipation. Or it can mean the establishment of new rules under a prevailing private property system" (Lai et al., 2016, p. 419, citations omitted).

The ability to create new property rights through contract or collective organization is often neglected by economists or misinterpreted as self-imposed regulation of public property without delving into the property right implications of those organizations and agreements (see Ostrom, 1990). Contracts and organizations create new systems of rules and liabilities and can connect various different types of property rights into a single instrument; they effectively modify property rights.

Ostrom (1990) and Libecap (2008), respectively, describe groundwater and oil and gas reservoirs as issues of open access to a commonly held resource, producing a tragedy of the commons (Hardin, 1968). Ostrom (1990) and Libecap (2008), however, do not recognize that property rights (exclusive control) were assigned with the application of the common law (water) and ferae naturae (oil and gas contexts) doctrines, which form the conditions of use, making these resources theoretical examples of private property, not common property regimes. As we have seen in previous chapters, when groundwater and hydrocarbons were considered private property, there have been issues with overproduction, waste, and inefficient competition as negative externalities due to the failures to internalize negative externalities in the design of the property rights (Libecap, 2008; Ostrom, 1990). In the hydrocarbon context, problems with the flow of hydrocarbons within the reservoir, interactions with gas and brackish water (transresources), and inefficient extraction methods could not be adequately addressed with regulation (Libecap, 2008; Olien & Olien, 2002). Instead, voluntary unitization agreements between overlying landowners created collective property rights to in situ hydrocarbon reserves out of the ferae naturae–derived property rights (Libecap, 2008). Other economists identify that unitization is a form of Coasian Bargaining, effectively reassigning property rights (Kaffine & Costello, 2011). While failing to recognize that unitization reassigned preexisting property rights through Coasian Bargaining, Libecap (2008) identifies that unitization agreements created (not modified) production rights that were far superior in addressing negative externalities, waste, and hydrocarbon transresources in the reservoir.

Likewise, the regulation of groundwater has failed to address the negative externalities associated with overconsumption, contamination, efficient use, and environmental damages. Ostrom (1990) identified that common-pool resource organizations (CPROs) allowed groundwater users to self-impose rules that improved the use and management of groundwater, essentially using water districts as a vessel to reform the property rights for groundwater. Like Libecap's (2008) mischaracterization of hydrocarbon reservoirs, Ostrom (1990) mischaracterizes groundwater as a common property, while the recognition of groundwater rights makes these rights private. The negative externalities of groundwater rights, as discussed earlier, are associated with the lack of internalizing the effects of groundwater use. Ostrom's (1990) solution, CPROs, effectively redefines the property rights as collective rights, with groundwater users each holding a share in the resource and participation in groundwater governance. These groundwater CPROs, however, fail to incorporate aquifer transresources into their governance system, unlike the unitization of hydrocarbon reservoirs, which includes gas, water, salt, storage spaces, and subsidence. Like unitization, these organizations have faced threats of anti-trust litigation, but may survive these challenges as a socially and environmentally beneficial form of collaboration (Scott, 2016).

While the debate over the categorization of groundwater rights as private or public will continue, groups have already begun efforts to reform the relationships and rights to groundwater. While unitization takes the form of a voluntary contract, these cutting-edge groundwater governance innovations use intergovernmental agreements, water districts, and associations to redefine, modify, and sometimes collectivize water rights to internalize negative effects of groundwater consumption. However, few examples have achieved aquifer governance by explicitly incorporating transresources or explicitly creating a system of collective property rights in an aquifer's transresources. To see the kind of governance success experienced in oil reservoirs, governance of the aquifer must be included, as well as the groundwater. The next section reviews a variety of examples of groundwater governance, showing the limitations for their use in addressing aquifer governance problems.

3.4 A Comparative Analysis of Examples of Groundwater and Aquifer Governance

The following examples of cutting-edge legislation, water districts, agreements, and associations provide an overview of the current approaches to dealing with the disconnection between water rights and negative externalities. Some of these examples employ a public right approach, mainly using regulations to solve the resource consumption issues. Others use private or collective rights approaches that challenge the dominance of top-down regulation. However, each of these examples fails to fully embrace aquifer governance, either using management approaches or only focusing on groundwater. The comparative analysis includes oil and gas unitization agreements and their aquifer counterparts, collective aquifer governance agreements.

3.4.1 The Traditional Groundwater Management Approach

For the sake of comparison, the traditional top-down model of groundwater management provides a baseline approach that places groundwater and aquifer governance in context. The traditional approach to groundwater management utilizes a state agency that determines the rights and obligations between individual water users, while also implementing statutory restrictions on a basin-wide level (Getches et al., 2015). Some states allow for the state agency to declare a special or critical area, subject to more restrictive rules, when groundwater depletions reach a certain level, often a safe or optimum yield of the aquifer (Getches et al., 2015). The traditional approach treats groundwater as a public good, subject to private use only through regulation. Ostrom (1990) calls this approach the "leviathan" approach, where state authority is used to force groundwater users to improve resource use patterns. Some attempts to apply the traditional approach have exposed states to litigation, where groundwater users claim the regulations "took" their property rights to the groundwater, requiring compensation (Newman, 2012). The traditional approach neither implements voluntary, negotiated, collective governance of groundwater nor incorporates transresources, including privately held storage spaces in the aquifer.

3.4.2 California's Sustainable Groundwater Management Act of 2014

Nearly fifty years after Elinor Ostrom's pioneering dissertation on groundwater dilemmas in southern California was published, and after experiencing severe groundwater depletions (Ostrom, 1965), California passed the landmark Sustainable Groundwater Management Act (SGMA) in 2014. This legislation creates a framework harnessing local governments, land-use agencies, and water districts called "sustainable groundwater agencies" (SGAs), which develop "sustainable groundwater plans" (SGPs) (Leahy, 2016). These SGPs must eliminate undesirable effects of groundwater depletions, including subsidence and water-quality degradation, and must reduce consumption levels to achieve sustainable groundwater use by 2040 (Leahy, 2016). If the SGAs fail to meet their statutory deadlines or the SGPs are insufficient, the state may impose its own interim plan until a local SGP is developed (Leahy, 2016). SGAs are authorized to impose pumping fees, develop recharge facilities, and require well registration (Leahy, 2016). However, the legislation states that the SGMA makes no alterations to groundwater rights in the state, potentially creating hurdles to actually implementing and enforcing the SGPs. The SGMA is a landmark legislation for locally governed groundwater governance, but attempts to address transresource problems (water quality, subsidence, saltwater intrusion) through groundwater regulation alone.

3.4.3 Innovative Water Districts, Agreements, and Associations

The traditional role of the water district can be envisioned as a vessel for groundwater governance. Water districts are authorized by state laws, allowing groups of water users to associate and gain some municipal powers, like eminent domain, retain real estate, and hold elections (Getches et al., 2015). During formation, a water district may force reluctant users into the district after enough users elect to form a district (Getches et al., 2015). These powers provide fertile ground for groundwater governance solutions. The following examples show how water districts and similar organizations can be used for groundwater governance to collectively manage groundwater, but can fail to achieve aquifer governance by failing to incorporate other transresources into the governance structure.

3.4.3.1 Rio Grande Water Conservation Subdistrict No. 1

Groundwater users in the San Luis Valley of Colorado voted to form the Rio Grande Water Conservation Subdistrict No. 1 in 2006, allowing the groundwater users to collectively govern their groundwater rights. Groundwater use in the San Luis Valley was intercepting surface water destined for downstream surface water users in neighboring states, potentially violating the Rio Grande Compact (Cody et al., 2015). After lobbying the state legislature, a law was passed that prevented state regulation of groundwater rights when a subdistrict implemented a management plan (Cody et al., 2015). The plan adopted by the Rio Grande subdistrict imposed a pumping fee and subsidized the fallowing of land to reduce groundwater consumption, which eventually led to a 32 percent decrease in groundwater pumping and the recovery of 250,000 acre-feet since 2013 (Bonini, 2018; Smith et al., 2017). Traditional state regulation of water rights was avoided by collectively bearing the burden of groundwater reductions and self-imposed abstraction fees.

3.4.3.2 Local Enhanced Management Areas

Local enhanced management areas (LEMA) allow groundwater users in Kansas to form their own groundwater management plans, rather than be subject to traditional state regulation (Peck et al., 2019). Kansas has adopted a regulatory system similar to Oregon, where regions with significant groundwater depletions may be designated as an "intensive groundwater use control area" (IGUCA) (Griggs, 2014). The unpopularity of the state-mandated plans with groundwater users resulted in new legislation allowing groundwater users to form their own management plans, creating a LEMA after its approval by the state authorities (Griggs, 2014). The plan, however, is enforced by the state authorities, as under an IGUCA procedure (Bonini, 2018; Griggs, 2014). The groundwater users may only limit the consideration of state authorities to the proposed plan, not manage its implementation. Under the LEMA statutes, only two groups of groundwater users formed LEMAs (Bonini, 2018). The LEMA approach, while an incomplete divergence from traditional state enforcement, does allow groundwater users to collaboratively develop the management plan determining groundwater use, providing an example of groundwater governance.

3.4.3.3 Umatilla Basin Water Commission and Intergovernmental Agreement

The Umatilla Basin Water Commission was created under an intergovernmental agreement between a water district, two counties, and Native American tribes in Oregon (Pagel, 2011). The commission was the result of decades of increased state regulation under a "critical groundwater area" designation, severely restricting groundwater use and development (Schroeder, 2016). The intergovernmental agreement allowed the coordination of the various participants, collectively using their administrative powers to implement a potential aquifer recharge project (Pagel, 2011). After an investigation found that the project would not have the intended hydrological effects, however, combined with a state requirement dedicating 25 percent of the water to environmental benefits, the project was cancelled (Pagel, 2016). Although the recharge project eventually proved unsuccessful, the Umatilla commission provides another example of a novel attempt to collectively govern a groundwater source.

3.4.3.4 Escalante Valley Water Users Association

The Escalante Valley Water Users Association (EVWUA) is not a water district but a nonprofit corporation for the sole purpose of pooling groundwater rights in the Escalante Valley of Utah. After groundwater levels decreased in the valley over a number of years, state regulators threatened to begin regulation of wells by priority, potentially forcing a 90 percent reduction in groundwater use overall, and leaving many groundwater users unable to access any water (Jarvis, 2011). In response, groundwater users formed the EVWUA in the hope that a coordinated effort could prevent such drastic reductions. In 2012, the State Engineer adopted the Beryl Enterprise Management Plan, which allows a voluntary agreement between groundwater users to proportionally bear the reductions as a group, rather than be subject to the priority system, thus treating the whole groundwater basin as a single unit (Jarvis, 2011; State of Utah Department of Natural Resources, 2012). The plan collectively governs the groundwater by converting the individual water rights into a shared right of access, making the EVWUA an example of a collective approach to groundwater governance, although it fails to incorporate the other aquifer transresources.

3.4.3.5 Model Memphis Sands Interstate Compact

Mondo (2018) developed a model interstate compact for the Memphis Sands Aquifer, which spans the subterranean boundary between Mississippi and Tennessee. The Model Memphis Sands Interstate Compact (MMSIC) incorporates participatory and adaptive governance of the aquifer through an agreement between the two states. A joint committee composed of agencies and entities from each state would direct the management of the shared groundwater resources (Mondo, 2018). Public outreach makes data available to the general public, but the joint committee determines the methods of maximizing the long-term benefits of the groundwater system (Mondo, 2018). Water quality is included in the planning conducted by the joint committee, but no other transresources are incorporated into the model compact.

3.4.4 International Examples: Morocco Contracts, Bellagio Draft Treaty, and Draft Law of Transboundary Aquifers

Two international examples move from groundwater management to aquifer management by incorporating transresources directly into the agreements but fail to achieve aquifer governance.

3.4.4.1 Morocco Water Contracts

Morocco began using contracts to gather groundwater users into a single management framework (Closas & Villholth, 2016). These contracts were nonbinding and were designed to reduce resource depletion. The contracts were negotiated at the local level and signed by the government and groundwater users and between the users themselves, and they included various local organizations

(Closas & Villholth, 2016). Groundwater users participated in the development of the means and goals of the contract frameworks using collaborative approaches (Closas & Villholth, 2016). The contracts were limited by conflicts over the issues and goals of the frameworks. State agencies believed that groundwater was being overconsumed, while local users believed the main issue was water distribution (Closas & Villholth, 2016). Small users, however, felt the contracts would unfairly benefit larger users (Closas & Villholth, 2016). The resulting contract for the Souss Aquifer limited new agricultural development, improved irrigation efficiency, imposed pumping fees, developed additional surface irrigation potential, and created research programs (Closas & Villholth, 2016). These voluntary agreements incorporated collaborative governance techniques and included some transresources (surface water), but were ultimately challenged by inconsistent goals and uneven sharing of costs and benefits between small and large users.

3.4.4.2 Bellagio Draft Treaty and Draft Law of Transboundary Aquifers

The Bellagio Draft Treaty between Mexico and the United States includes transresources like aquifer pore spaces, groundwater quality, surface interactions, land subsidence, and aims to protect the "underground environment" (Hayton & Utton, 1989). However, the Bellagio Draft Treaty uses an interstate commission composed of agents appointed by the governments to manage the aquifer, and it does not include individual groundwater users. The interstate commission regulates both surface and groundwater in the region, which prevents duplication and ensures uniform management goals of the interconnected resources.

Likewise, aquifer transresources have been identified as important aspects of proposed international law. The DLoTA specifically incorporates transresources into the management of transboundary aquifers (McIntyre, 2011). The draft articles incorporate heat, waste storage, pollution, carbon sequestration (identified in supporting commentary), and minerals contained within the "aquifer system" (International Law Commission, 2008). The draft articles also explicitly recognize that some aquifers may not be recharged, providing for the management of "fossil aquifers" that receive negligible recharge.

The DLoTA creates a duty for transboundary states to develop joint management plans (International Law Commission, 2008). However, like the Bellagio Draft Treaty, individual groundwater users have no role in the negotiation or development of the management plans, making the DLoTA an aquifer management system that fails to achieve aquifer governance.

3.4.4.3 Agua Caliente Litigation and Pore Space Property Rights

Transresources may be forced into the groundwater governance system within the United States by court order. In a lawsuit brought by the Agua Caliente

Band of the Coachella Indians, the Tribe claims to possess reserved groundwater rights within the aquifer below their reservation lands (*Agua Caliente Band of Cahuilla Indians* v. *Coachella Valley Water District* et al., 2018). On appeal, the Tribe successfully defended the finding that tribal groundwater rights are included as part of the Tribe's rights to water. Additionally, the Tribe claims that the right includes a specific right to water quality and possession of the pore spaces within the aquifer.

The motivation for the claim comes from alleged filling and contamination of the aquifer by neighboring artificial recharge projects conducted by the Coachella Valley Water District (CVWD) and Desert Water Agency (DWA). The Tribe has presented evidence of three states (Montana, Wyoming, and North Dakota) that statutorily recognize private property rights in pore spaces below landowner's surface estate (*Agua Caliente Band of Cahuilla Indians* v. *Coachella Valley Water District* et al., 2018). While the outcome of the litigation is still pending, the potential property interests in pore spaces and groundwater quality could make groundwater governance (or management) inadequate when challenged with many overlying property interests and conflicting aquifer uses. Artificial recharge projects could face a multitude of trespass or conversion claims.

This litigation could potentially unravel the narrow focus on groundwater within the United States, forcing a new model of aquifer governance that incorporates private rights to transresources in aquifer systems and requires voluntary participation.

3.5 Comparing the Examples

Unitization provides a real-world example of transresource governance of the subsurface environment. The preceding section outlined the emerging and cutting-edge examples of groundwater management and governance, and also their limitations in achieving aquifer governance. This section compares the examples of the previous section, showing the common aspects and placing them in categories of groundwater management, groundwater governance, aquifer management, and aquifer governance. The section introduces unitization agreements, as commonly used in the oil and gas industry, and incorporates them by analogy into the comparative analysis. The section also incorporates the system analysis discussed earlier, showing the examples' success in linking the social-legal system and the transresource system.

These examples are compared using the resource-governance scale. As seen in Figure 3.4, each scale contains characteristics, enabling characterization of each example into the paradigms described earlier. Specific components of the resource scale include groundwater, pore spaces, and other resources and the incorporation

	Traditional approach	SGMA	Model memphis sands compact (Mondo 2018)	Innovative water districts and agreements	Morocco contracts	Bellagio draft treaty and DLoTA	Agua caliente litigation	Collective aquifer governance contracts	Oil and gas unitization
Resource Scale Groundwater?	X	X	X	X	X	X	X	X	X
Pore Spaces?		(X)		(X)		X	X	X	X
Other Transresources?			(X)			X	X	X	X
Adaptive Management?	(X)	(X)	X					X	X
Social Scale Public, Private, or Collective Rights?	Public	Private	Public	Collective	Public	Public	?	Collective	Collective
Collaborative?				(X)	X			X	(X)
(Re)Negotiated?		(X)			X			X	X

Figure 3.4 Comparison of cutting-edge groundwater approaches

Aquifer governance

Aquifer management

Groundwater governance

Groundwater management

of adaptive management, making key distinctions between groundwater and aquifer paradigms. The social scale includes legal interpretation of rights (public, private, or collective), the collaborative aspects of the governance approach (voluntary, user participation), and the negotiation and renegotiation of the management decisions. The social scale components provide the key difference between management and governance paradigms.

The final portions of this chapter will introduce oil and gas unitization and compare the approach to the examples, analyzing unitization within the framework of social and resource scales. The sections will identify the specific characteristics that relate to the groundwater-aquifer context, showing that unitization can be used as a model for aquifer governance.

3.5.1 Groundwater Management Examples

Groundwater management is the traditional approach used for the allocation of the water resources within an aquifer. As shown in the inner solid square in Figure 3.4, three examples fall into this paradigm: the "Traditional Approach" (see Chapter 8 for a detailed discussion, using the potential Harney Valley critical area as an example), the SGMA, and the Model Memphis Sands Interstate Compact. These examples show attempts to manage problems like groundwater depletion, subsidence, groundwater quality degradation, and surface water interference indirectly through state regulation. They fall into the regulation-controversy feedback, where reductions in groundwater allocations are attacked as a public taking of private property or are resisted by portions of the public (Brown, 2015; Owen, 2013). In the MMSIC example, the interstate compact proposed by Mondo (2018) uses a public law governance approach, while identifying that groundwater is owned privately in Tennessee; this potentially creates a similar regulation-controversy feedback, because actions taken by the joint committee would likely reduce allocations for Tennessee groundwater pumpers, taking private rights.

Transresources are only indirectly incorporated into these examples. The SGMA partially incorporates pore spaces into its planning regime, specifically addressing subsidence as an undesirable result of groundwater consumption. The MMSIC includes groundwater quality and a distinction between the aquifer and groundwater in supporting materials, but no other transresources (Mondo, 2018). The first two examples somewhat incorporate adaptive management, allowing for removal of a critical designation or amendments to plans developed under the SGMA. The MMSIC explicitly incorporates adaptive management and mandatory revision of plans at regular intervals. The legal approach of the SGMA differs from the Oregon's Umatilla Basin critical groundwater area approach, in that it treats groundwater as a public resource, using top-down requirements regarding the imposition of regulations to reduce groundwater consumption. However, the SGMA claims to make no changes to groundwater rights in California, treating the right to extract groundwater as a private right. The MMSIC's joint committee implicitly takes a public ownership approach, giving the committee power to oversee groundwater use in the aquifer boundary without regard to individual water rights.

Nonetheless, these examples do not incorporate collaborative, participatory approaches or other transresources, making these expressions of the traditional regulatory approach. Critical groundwater areas are designated by a state agency, with minimal requirements for public participation. The SGMA requires the development of SGP by local agencies, which may not include any local participation or collaboration. The MMSIC implements a joint committee of agencies and groups in charge of groundwater but reserves no places for groundwater users within the aquifer boundary. Because these examples directly incorporate neither transresources nor governance principles, these fall into the groundwater management paradigm.

3.5.2 Groundwater Governance Examples

The groundwater governance paradigm incorporates participatory, collaborative processes into the decision-making process, but does not directly embrace aquifer transresources. The collection of innovative water districts, agreements, and associations, along with the examples of the Morocco contracts, fall under groundwater governance, as shown in larger dashed lines in Figure 3.4. The innovative water districts used current laws in novel ways to respond to the traditional top-down regulation of aquifers, creating a more collective rights–based legal regime. Individuals within these water districts shared the burdens and costs of groundwater consumption collectively, making decisions locally in response to physical and legally imposed groundwater unavailability. These water districts only indirectly incorporate pore spaces, assuming the ability to recharge and store water within the aquifer as an implied right. These water districts are only partially collaborative, allowing groundwater users greater access to locally controlled boards and directors, but not directly involving users within the decision-making process. Water districts are not negotiated or renegotiated, because the organization becomes permanent once created, and unwilling parties may be forced into the district at its inception.

The Umatilla commission and EVWUA both focused their efforts on groundwater and assumed the use of pore spaces for recharge, or simply ignored potential property rights to storage within the aquifer. The Umatilla commission was indirectly collaborative in relation to individual groundwater users, with their interests represented as members of the commission without direct participation. The EVWUA was collaborative, since participation was voluntary and initially negotiated, but without the ability to renegotiate at a later time, due to the strict conditions imposed by the state agencies. Like the earlier water district examples, the Umatilla commission and EVWUA used a collective rights approach, where burdens and benefits are distributed as shares to individual groundwater users. The Morocco contracts exemplify a collaborative, negotiated governance approach.

While the contracts did claim to include water quality considerations, the eventual contracts did not directly include transresources in their agreements. Individual users negotiated with state agencies and other groundwater users to implement water projects and limit consumption. The conditions within the contracts were negotiated on an individual basis, and the agencies allowed voluntary compliance with the contractual terms. However, these contracts used a public rights approach, giving agencies the power to authorize and charge fees for groundwater use.

The use of nontraditional agencies to implement policies and move control from top-down regulation to bottom-up governance sets these examples apart from the previous paradigm. These examples of groundwater governance include the collaborative, participatory, and voluntary approach, but fail to address the aquifer's critical role. The following section explores examples of aquifer management where transresources are addressed, but minimal collaborative or negotiated governance occurs.

3.5.3 Aquifer Management Examples

Aquifer management builds on groundwater management by directly addressing aquifer transresources in the management of an aquifer system. As shown in the smaller dashed lines in Figure 3.4, three examples can be considered as examples of aquifer management: the Bellagio Draft Treaty, DLoTA, and the Agua Caliente Litigation. The Bellagio Draft Treaty specifically mentions aquifer storage, heat, waste disposal, and other transresources within the text of the document, incorporating their roles in the aquifer planning process. Likewise, the DLoTA includes these transresources as components of the aquifer for planning purposes. However, neither the Bellagio Draft Treaty nor the DLoTA incorporates collaborative, voluntary, or participatory governance strategies. Because both of these international legal instruments govern the acts of nations, they implicitly take a public law approach by addressing governance through national territory and sovereignty, not individual and human rights.

The Agua Caliente litigation may also be categorized as an aquifer management scheme. The litigation is similar to the Bellagio Draft Treaty and DoLTA in that separate rights between nations are being established. Although tribes are considered domestic dependent nations, the legal system has not addressed the rights granted to each party in the litigation. Additionally, the litigation centers on transresources: the possession of pore spaces and rights to water quality. The outcome of the litigation is unknown, but will likely not create collaborative, participatory governance. Should the DWA and CVWD succeed in the litigation, the aquifer would be treated as a public resource available to all. If the Tribe is successful, the aquifer transresources will be privately held, creating the regulation-controversy feedback or trespass-conversion scenarios discussed earlier. The court could find a collective right held by all parties concerned with the jointly held aquifer, but this position has not been argued by either party. The individual groundwater users within the DWA and CVWD will only be indirectly participating in the eventual permanent solution to the conflict. Negotiation of a settlement, if it occurs, will only involve the parties to the litigation and will likely not be renegotiated, absent additional conflicts and litigation.

None of the examples in this section directly bring together aquifer transresources into a governance system using collaborative, participatory, or voluntary approaches. Like the traditional regulatory paradigm, these approaches implement primarily top-down management and do not include individual groundwater users in the management of the aquifer. The next section introduces an allegorical example of aquifer governance: the unitization agreement used in the oil and gas industry.

3.5.4 Aquifer Governance? Oil and Gas Unitization Agreements

At this time, few examples of a transresource-based collaborative governance exist with groundwater as the primary concern. Instead, unitization agreements can be used as a model governance system that can be applied to aquifers. The model of Collective Aquifer Governance Agreements, as shown in Figure 3.4, would fully embrace transresources while also incorporating collaborative, participatory, and voluntary governance approaches. These agreements represent many of the positive aspects of the previous examples in a single governance approach. As can be seen in Figure 3.5, aquifer unitization encompasses all the aspects of the other four approaches we have outlined.

Unitization includes various reservoir transresources like oil, gases, pressure, water quantity, water quality, subsidence, heat, and any other relevant resource within the reservoir boundary. Unitization converts private rights in oil and gas reservoirs into collective shares, similar to the way water

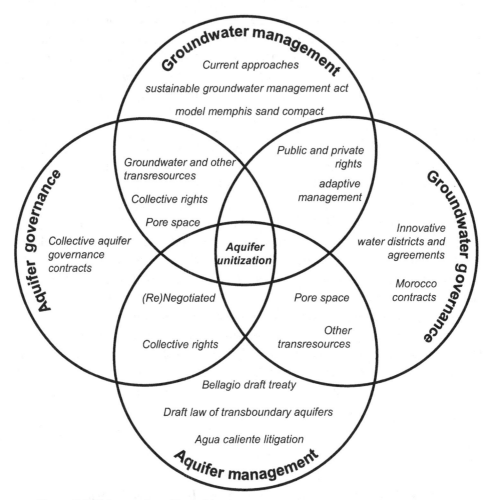

Figure 3.5 Spheres of aquifer unitization

districts spread the benefits and costs of aquifer development. Additionally, the agreements are often voluntary (with exceptions) and are renegotiated at regular intervals or when resource conditions substantially change. Chapter 4 introduces the principles of unitization and applies these principles to a potential Collective Aquifer Governance Agreement.

4

Unitization and Collective Aquifer Governance Agreements

Collective aquifer governance agreements (CAGAs) have the potential of transforming management into governance and expanding the scope from groundwater to the aquifer. Unitization has the potential to bring the transresource system (see Figure 3.1) together with the social-legal system (see Figure 3.2) using a collective property rights system (see Figure 3.3).

Unitization agreements create a new system that links the social and resource domains. In the next section, unitization principles will be outlined, showing how the unitization approach creates a new system of rights, resources, and governance within a reservoir. The following section of the chapter will discuss the specific components of unitization agreements and apply them to the groundwater aquifer context through CAGAs. A timeline for CAGAs will be outlined, providing the procedural chronology used to form an effective CAGA. Specific suggestions for CAGA components will be discussed on the basis of the different resource conditions between the two contexts, including the transresource shares, plans, and effects of redetermination.

4.1 Unitization Agreements in Oil and Gas: A Systems Approach

Unitization has become the default approach used in the oil and gas industry for the operation and governance of a hydrocarbon reservoir since the late nineteenth and early twentieth centuries. The reasons for the development of the unitization approach are similar to the issues faced by aquifers today: ineffective regulatory restrictions hampered by controversy, takings litigation, unplanned extraction by many users, insufficient knowledge of the resource, and waste. The answer came in a multifaceted agreement applying new principles to reservoir governance that incorporated social and resource domains, using collective shares in the resource. As seen in Figure 4.1, unitization agreements form the center of the unitization agreement system, linking rights, users, transresources, and legal requirements.

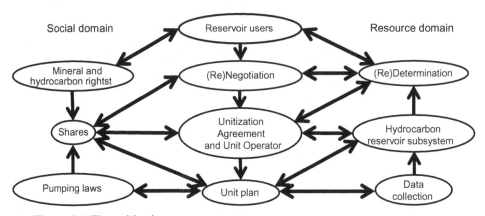

Figure 4.1 The unitization agreement system

The unitization agreement system begins with the reservoir users negotiating an agreement that converts their own mineral and hydrocarbon rights into collective shares within the reservoir within the social domain, linking the social domain with the resource domain. As shown in Figure 4.1, a governance system is created for the reservoir. The timeline of unitization moves from exploration and discovery, to determination of shares and agreement negotiation, to production, and then to abandonment of the field (Asmus & Weaver, 2006; Easo, 2014; Igiehon, 2014; Lowe, 2014). Exploration involves acquiring rights and permits to drill, gaining exclusive rights in the area, and finding profitable reservoirs. When a productive reservoir is found, the reservoir users negotiate shares of the commonly held reservoir and the agreement that determines the unit operator conducting production (Asmus & Weaver, 2006). Production is the main oil-producing phase of the life of an oilfield, often managed by a unitization agreement (Asmus & Weaver, 2006). Abandonment occurs when the reservoir is depleted and closing operations are conducted, potentially allowing new uses of the reservoir (Igiehon, 2014).

The unitization agreement stands between the social and resource domains. Reservoir users negotiate the terms of the unitization agreement, selecting a unit operator, and implementing a unit plan (Figure 4.1). The reservoir users play the central role in the governance system, with changes in the system eventually resulting in their direct involvement and negotiation. The unit operator is an entity that carries out the terms of the unitization agreement but does not necessarily play a role in negotiation (unless they are also a reservoir user) (Onorato, 1977; Shaw, 1996). The unit operator creates the unit plan, selecting the most efficient means, and conducts the production in the reservoir (Shaw, 1996). Unit operators may be removed and replaced by a vote of the reservoir users for no reason, willful misconduct, or violations of the unitization agreement (Asmus & Weaver, 2006;

Shaw, 1996). The unit operator is the primary entity carrying out the operation of the unitization agreement, bridging the social and resource domains.

The social domain (Figure 4.1, left side) is composed of the rights and laws that govern hydrocarbon resources. Rights to hydrocarbon extraction are based on private property, specifically the rule of capture, allowing reservoir owners to extract as much as physically possible on their land without regard to other users. These property rights to hydrocarbons and minerals come from the same legal doctrines of ferae naturae and property rights discussed in earlier chapters (Weaver, 1986). Courts later restricted these rights to correlative rights, giving each pumper a "fair share" of the hydrocarbons below their land and preventing waste of the resources (Kramer & Anderson, 2005). The limitations on property rights places oil and gas resources in the private property regime, not a common property system (a system of no private rights), as often described by economists (Barbanell, 2001; Libecap, 2008; Ostrom, 1990). Pumping laws are limited to well spacing, well construction, maximum pumping rates, and waste limits (for gases) (Kramer & Anderson, 2005; Olien & Olien, 2002; Weaver, 1986).

Frequently, exceptions have been made to these laws, under threat of taking claims, which undermined regulation, similar to some current challenges to groundwater laws (Olien & Olien, 2002). By contract, pumping rights are converted into shares of oil, gas, water, storage spaces, and other reservoir constituents representing the amount of the resource present in situ or below the user's land (Asmus & Weaver, 2006). Shares are determined using multiple methods, typically representing an estimation of the recoverable volumes of hydrocarbons within the reservoir (David, 1996). These shares combine the rights held by the users with the legal requirements imposed on the oil and gas industry. These shares replace the rule of capture rights (associated with negative externalities and waste) with in situ rights by contract. The conversion to a system of shares of a common resource represents the conversion of private property into collective property rights. The allocation of shares between reservoir users determines the distribution of benefits, costs, and the voting power of each user in the agreement (Asmus & Weaver, 2006; Beggs & Stockdale, 2014; David, 1996). The combination of the property rights giving access to reservoir resources, the laws that regulate their use, and the shares within the unitization agreement form the social domain in the unitization system.

The resource domain (Figure 4.1, right side) represents the physical and transresource conditions in a reservoir. During the negotiation of the agreement, an initial distribution of shares is made based on available information (Asmus & Weaver, 2006). This determination is based on the physical conditions within the reservoir, but is redetermined on a regular basis or when new information (like a new reservoir boundary) is discovered that would substantially change the distribution of shares (Beggs & Stockdale, 2014; David, 1996). The unit plan and unit operator must collect production information, and this information is used to redetermine the shares held by the reservoir users. The redetermination is conducted using a method negotiated by the reservoir users in the same fashion as the original distribution, rebalancing the shares to match the actual resource conditions in the reservoir based on the new data. The redetermination process returns the unitization system to the reservoir users.

The unitization system creates a positive feedback linking the social domain's system of shares (costs and benefits) with the resource domain (actual resource conditions). The emergent property of this system has been a robust, reservoir governance system based on the conversion of private property interests into a collective shares system, regularly redetermined based on new information, distributing costs, and benefits of the reservoir to each reservoir user. A similar approach could be used to govern aquifers.

4.2 Collective Aquifer Governance Agreement Systems

A suggested collective aquifer contract would mirror many of the components of a unitization agreement. As depicted in Figure 4.2, many of the functional elements of unitization agreements would be present in a CAGA. With the complexity of the modern water law system, the social-legal system creates a subsystem within the CAGA governance system. Likewise, the additional complexity of the aquifer transresource system is reflected in the CAGA governance system as a subsystem.

In the CAGA governance system, aquifer users form the main decision-making and negotiating body. While water is likely the primary concern in the

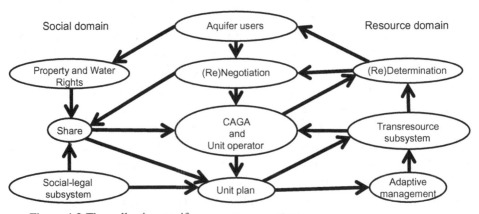

Figure 4.2 The collective aquifer governance system

aquifer, any aquifer transresource could form the basis for a CAGA. This book assumes water as the central focus of a CAGA for simplicity and application to many regions. These users hold water and property rights within the aquifer region and negotiate the form of the shares that make up the voting power within the CAGA. The creation of shares internalizes the externalities of groundwater (or other transresource) uses, reassigning new property rights by contract, as discussed later (Lai, Davies, & Lorne, 2016). These shares are also partially reflected by the effects of the social-legal subsystem providing regulation and public oversight of the CAGA. The CAGA integrates the regulations, laws, and environmental oversight of state agencies into the shares and the overall CAGA governance system.

The aquifer users select a unit operator that develops and conducts the unit plan, the coordinated use and management of the aquifer's transresources (Figure 4.2). The unit plan determines the objective and means of achieving the governance of the aquifer. For example, a unit plan may incorporate a recharge facility, pipelines from more-productive to less-productive portions of the aquifer, or surface-flow augmentations for environmental benefits. A unit plan may also determine the conversion and transfer of shares within a water marketing and banking system.

Adaptive management in the CAGA takes the place of data collection under unitization agreements. As the CAGA is implemented, the unit operator will collect information about the aquifer and adapt the unit plan, with consent of the aquifer users, to achieve the goals set out in the unit plan (Craig & Ruhl, 2014). Likewise, the adaptive management and information gathering may trigger a redetermination of the aquifer users' shares, and further renegotiation of the CAGA or unit plan. The CAGA governance system includes voluntary, participatory, adaptive governance and directly connects the social and resource domains.

In the following sections, the chapter reviews the components of a CAGA. Because these contracts would be specific to the community creating them, a variety of options and alterations could be made to the broad outline presented in the Model Collective Aquifer Governance Agreement later in this book (see Appendix). Like unitization agreements, these contracts would be individualized by the needs specific to the community, physical conditions, and legal requirements in a jurisdiction. The unitization approach to aquifer governance would involve many of the same negotiations and compromises within the broad framework of unitization.

The section begins with a description of the potential timeline of aquifer unitization. Next, the potential components of an aquifer unitization agreement are reviewed. While this section presumes that water is the central transresource, any aquifer transresource could form the keystone resource for a CAGA.

4.3 Timeline for a Collective Aquifer Governance Agreement

Groundwater development has progressed in a much different way than that of oil and gas. While oil and gas discoveries are limited to select areas of the world, groundwater is everywhere. The timeline for groundwater governance and development has occurred over centuries, not the decades seen in oil and gas in the early twentieth century. While many aquifers have been discovered, developed, and have considerable production, there are still more aquifers yet to be discovered. Today's aquifers exist somewhere between underdevelopment, overdevelopment, or unplanned mining. The renewability or nonrenewability of groundwater changes the dynamics in aquifer development in an aquifer unitization agreement. The groundwater development and agreement timeline exist at different stages depending on the location and physical conditions of the resource and the point of view of the reader. A depleted aquifer may be abandoned for some uses, or may be a great place to explore for storage using an aquifer storage and recovery (ASR) system.

4.3.1 *Phase 1: Exploration and Discovery*

Groundwater aquifers have been explored on every continent, basin, and region on the globe. As discussed in previous sections, groundwater has been used for human and agricultural uses for thousands of years. Unlike the rapid expansion and need for oil, water's exploration history has had millennia to develop, through myth and methodology, and to assign various kinds of rights to its use. With oil and gas, exploration for reservoirs is encouraged by the lease, license, or concession system, granting well developers with exclusive rights to the benefits from their exploration activities. Groundwater, on the other hand, has already issued many of the rights to its exploration and use. Like oil and gas, groundwater rights within the United States are tied to property ownership and permitting systems. Landowners are granted the exclusive right to drill wells for groundwater within their land and bear the risks and rewards for their exploration.

However, even exploration has been limited to finding sources of groundwater, typically not aquifers themselves. Exploration for aquifers would include the potential for storage of producible quantities of water within the pore spaces and fractures of the earth. As discussed later, when aquifers reach a point of depletion (especially fossil aquifers), these pore spaces have the potential for storage and secondary uses. Secondary uses would come after the end of the primary use of the aquifer: to primarily produce groundwater. Aquifer storage and recovery, water marketing and banking, water quality improvement, and temperature regulation may be secondary uses of aquifers. Instead of discovering new sources of groundwater, this phase explores and discovers new uses and goals of existing aquifers or entirely new ones.

4.3.2 Phase 2: Determination of Rights, Shares, and Governance Structures

Before an aquifer can be unitized, a pre-unitization agreement could form the initial structure that would form the final unitization agreement. A preliminary aquifer unit committee (or "pre-unit committee") would be formed from the interested aquifer users in a community. These could be representatives from water districts; local elected officials; leaders of nonprofit organizations, environmental groups, or soil and water districts; or a collection of individual aquifer users in a community. The pre-unitization committee would select a unit operator to conduct the early development of the unit. The unit operator could be a geotechnical consulting agency, a participating water district, a nonprofit organization, or some other technically capable member of the pre-unit committee. The pre-unit agreement would include protection of confidential information and data provided to the unit operator to assist in the research of the aquifer.

The aquifer unit operator would begin collecting the data available to determine the preliminary shares in the aquifer. Each owner of a groundwater well (or potential surface water right) could be issued equity interests based on their current water rights. These preliminary equity interests would determine the voting rights, benefits, and costs of participating in the aquifer unit committee. The pre-unit committee could fund additional hydrogeological studies to determine the inflows, outflows, geologic information, and storage potential for the community. Funding comes proportionally from equity interests, grants, and other sources of revenue located by the unit operator. The preliminary unit interests are based on physical values, like land area, groundwater production, potential groundwater production, pore space volume or some other kind of physical aquifer characteristic. A negotiation of research costs, interests, and economics results in the final selection of the interest calculation method. The equity interests roughly reflect the status of the aquifer at the time of unitization.

The unit operator would develop a unit plan to efficiently optimize the aquifer to the goal decided by the pre-unit committee. The unit operator develops a plan using technical and economic information to maximize the social benefits of the aquifer and water resources, using the participants' unitized rights. The plan could include installing jointly owned ASR projects, well stimulation, recharge basins, applications for new jointly held water rights (available winter flows, for example), or irrigation technology improvements.

The pre-unit committee could approach non-participating parties and offer them participation within the unit. These nonparticipating parties could be offered equity interests, as determined by the pre-unit committee. As the number of participants grows, the next phase of negotiation would begin: to determine the goals, potential benefits, and rationalization scenario for the aquifer (Foster et al., 2003). The goal could be gradual recovery, stabilization, or controlled depletion, depending on the social and physical conditions of the aquifer.

4.3.3 Phase 3: Operation of the CAGA

Once the aquifer and watershed have been technically investigated and equity interests have been negotiated, the participants in the preliminary unit committee could sign the final aquifer unitization agreement. Because there is no compulsory aquifer unitization statute, each member voluntarily joins the unit because of the potential collective benefits it could bring. This agreement would lock in the equity interests (until the next redetermination) and begin the joint development of unit projects, like ASR or other water management schemes. For example, water efficiency improvements free water to be stored within the ASR system, increasing the collective storage within the aquifer. A market for the different types of equity interests could begin. The unit operator conducts the analysis of the aquifer as operations continue and gathers data owned jointly by the unit. Various redeterminations are conducted as the unit operator gathers more hydrogeological information, and equity shares are redistributed as decided by the unitization agreement.

4.3.4 Phase 4: Abandonment or Repurposing

Unlike oil and gas production, many aquifers may never be truly abandoned. A unitized unconfined aquifer may go through many redeterminations and improvements as technology and hydrogeologic knowledge improves. Climate change may also require regular redeterminations if natural recharge rates shift as precipitation and surface water availability shifts. Outright abandonment of many aquifers is unlikely in most cases.

For the case of fossil-confined aquifers, the groundwater may not be completely withdrawn according to a rationalization scenario (Foster et al., 2003) due to legal requirements of the jurisdiction. While the primary use of a fossil aquifer will be complete, a unit operator could potentially incorporate its secondary use as a storage reservoir for the unit. Aquifers that suffer from subsidence or inelastic compression, if allowed by law, may be abandoned after use by the unit, since the aquifer may never serve another water function. Abandonment of the unit may occur if the unit dissolves due to other circumstances.

Because unitization of aquifers would be entirely voluntary as a contract (absent new legislation akin to water district formation), the governance system could end due to internal conflict, social disruption, or legal attack. A CAGA abandonment agreement would be needed to close down CAGA operations. Because of the potential to fail, a fund or bond should be maintained in the CAGA to ensure that money remains available for this circumstance. This agreement would involve the costs of closing joint facilities, returning any data to its owner, and clarifying liabilities for individual participants.

The timeline for aquifer unitization agreements follows the phases seen in the oil and gas context, with some alterations. While oil and gas have a linear progression from discovery to abandonment, groundwater has been in a state of uncoordinated production since its discovery. Unitization agreements provide a process of converting the chaos of current groundwater management into a more productive and harmonious governance system. The next section describes the specific components of a CAGA and how these concepts could be applied to a collectively governed aquifer.

4.4 Recommendations for Components of a Collective Aquifer Governance Agreement

Many of the components of oil and gas unitization agreements can be directly applied in an aquifer unitization agreement, with some minor modifications. The specific goals and purposes of aquifer unitization will be considerably different from the oil and gas context. A rationalization scenario, or other purpose, will replace the efficient extraction as the ultimate goal (Foster et al., 2003). Likewise, the equity interests in a collective aquifer agreement will reflect different transresources within the aquifer, like storage, replenishment source, and other factors. Additionally, the timeline variations discussed earlier will influence the redetermination process, reflecting changes in scientific, legal, and policy goals. The following section describes the various components of an aquifer unitization agreement and the potential points of negotiation that would be decided by participants.

4.4.1 Purposes and Goals of the Unit

The purposes and goals of a collective aquifer governance agreement would be a combination of legal, environmental, economic factors, and social factors, like community, sustainability, and identity. The potential participants would determine the goals for the unit, which would eventually be enshrined in the CAGA. The purposes would likely reflect the general rationalization scenarios discussed by Foster et al. (2003) informed by current laws and regulations. The unit would determine the desired outcome: gradual depletion, stabilization, or managed recharge of the aquifer. Other goals could include development of community hydrothermal systems, bioremediation, groundwater contamination management, or the prevention of saltwater intrusion. The choice would be narrowed by the legal requirements available in the existing system of rights, laws, and policies. For example, the gradual depletion rationalization scenario may subject the unit to potential designation as a critical groundwater area, which would potentially hinder unit operations.

The three rationalization scenarios represent the general goals that could be used by a unit, but the unitization agreement could be as specific or general as the participants desire. For example, the motivation to unitize could be the community's desire to lift a critical-area designation, expand irrigation potential, provide resilience to climate change, improve groundwater quality, improve environmental flows in rivers (protecting an endangered species), restore water-table levels, or create a water bank in conjunction with a water market. The CAGA would make the goals clear for the unit operator as they begin designing systems for and managing the aquifer.

One critical difference between oil and gas unitization and collective aquifer governance is the physical purpose of the unit. For aquifers, the purpose would be the preservation, delivery, and use of groundwater by members of the unit, not the maximization of profits seen in the hydrocarbon context. In some cases, aquifer users may accept money in exchange for their access to groundwater. In other cases, aquifer users may need to withdraw groundwater to maintain their living. "Maximum efficiency" for groundwater may be the sustainable distribution of the groundwater in an aquifer, not extraction. The aquifer unit plan will look much different from an oil and gas unit plan.

4.4.2 Formation of the Unit

Because no compulsory CAGA statute exists, these agreements would be completely voluntary. The problem of free riders to the benefits of a unitized aquifer would exist, wherein neighboring users would benefit from the CAGA unit while not participating in the CAGA itself. Incentives to join the CAGA include the benefits of collectively sharing costs and risk, participation in water markets and banking, and access to technical information. Climate change, additional regulations, or the wish to expand the potential of the aquifer(s) could also provide the needed incentives to promote cooperation. These incentives are unlikely to completely eliminate the free-rider issues. A compulsory unitization statute, similar to the way typical water districts are formed, could ensure that nonparticipating parties receive a windfall from the joint development of aquifer facilities (Getches et al., 2015). Water districts are formed by a petition of a certain number of water right holders in a region and can force others into the district. A simultaneous creation of a CAGA and a water district could solve the issues of free riders, if permitted by law, creating a pseudo aquifer-unitization statute through the use of groundwater district statutes.

4.4.3 *Aquifer Unit Operator*

The aquifer unit operator will take the leading role in the development of the unit plan, overseeing its operation and determining the specific projects to meet the goal selected by the unitization agreement. Like in the oil and gas context, the unit operator could be a member of the unit committee or a third party without an interest in the agreement. The role of the unit operator is essentially a "nonprofit" role, since no additional benefit is gained from the position. If the unit operator is a member of the committee, the only benefit they would receive is the same proportional improvement in aquifer governance as any other interest holder. If a third party is selected, their expenses for operating the unit would come from the joint account as a unit cost. Potential organizations that could serve as an aquifer unit operator are water districts, law firms, nonprofit organizations, geotechnical consulting firms, academic institutions, institutes, or any other organization with the technical capacity to conduct aquifer studies, development, engineering, and governance. The unit operator could partner with other organizations if the unit operator does not have all the skills or capacities required for the position.

The unit operator develops the unit plan, fulfilling their obligation to determine the projects and strategies to achieve the goal outlined by the CAGA. For example, the unit plan may incorporate applying for new surface and groundwater rights, development of ASR and recharge basins, well development or stimulation, or in situ bioremediation (EPA, 2013).

4.4.4 *Aquifer Unit Committee*

The aquifer unit committee provides oversight and decision-making power within the CAGA governance system. The aquifer unit committee would be composed of representatives of the holders of equity interests in the unitization agreement. The voting power of each member would reflect the party's shares in the unit. The aquifer unit operator provides the participants with reports and data to enable the committee to make decisions for the unit. The participants and shareholders would make all major decisions on unit expenditures and project approvals, including the unit plan.

Unlike the handful of oil companies in oil and gas unitization with large percentages of the unit, an aquifer unitization agreement would likely have a multitude of small-interest holders in the aquifer unit. The voting mechanism could be provided using a participatory geographic information system (PGIS) software package (Mekonnen & Gorsevski, 2015). Users would be able to make siting and other decisions through the use of PGIS mapping and data collection systems. Additionally, the number of members of the unit committee would require innovative methods of voting. Cell phone applications could potentially be integrated with PGIS to allow for data collection from participants' wells for voting or for contacting the unit operator when issues arise (Kangas et al., 2015). These systems could also be used in the pre-CAGA phase, as the unit plan is developed, to improve the collaborative approach required by the CAGA governance system, empowering aquifer users in the governance of the unit (Kwaku Kyem, 2004). Because the CAGA governance system rests on the rights of the voluntary participants in the unit, the unit committee empowers the members to make decisions for the unit, the highest form of collaborative governance (Nyerges et al., 2006).

4.4.5 Redetermination

Because the equity interests are based on natural physical conditions in the aquifer, and the operation of the unit will generate new data, the initial distribution of interests in the collective aquifer may require periodic revision. As discussed more fully in later sections of this book, the redetermination (or concurrency assessment) could occur a few years after unit operations begin, by request of the unit committee, at specified intervals, at certain thresholds in the unit plan (like improvement in the water table), or after the discovery of data that significantly alters the distribution of equity interests. In fossil aquifers under a gradual depletion unit plan, redeterminations could occur infrequently. The expense may be too large to warrant extensive studies of the confined aquifer system. In recently developed aquifers, the lack of data could justify further redeterminations as operations continue. The unknown changes to precipitation due to climate change may require periodic redeterminations as the effects are increasingly felt. The trigger for redeterminations is another point of negotiation, balancing the costs and benefits with the goals of the unit.

4.4.6 Information and Adaptive Management

While much of the data for aquifers are publicly available, the data regarding the actual use of wells by individuals is predominantly privately held. A major component of the negotiation of a CAGA would be the exchange of any existing data between potential participants and the potential unit operator. This data is exchanged confidentially, improving the trust among aquifer users and the preliminary unit operator. After the CAGA is negotiated, any data collected by the unit operator is owned collectively by the aquifer users. Public knowledge of surface and groundwater use data could create conflict between members of the unit or entail legal ramifications for participants. The importance of confidentiality in data sharing should be an important aspect of the pre-unitization and unitization agreements.

The unit operator would incorporate adaptive management into the unit plan, allowing for changes based on new conditions, including regulations and climate change, to alter the unit operator's activities. The decision to redetermine shares, adjust the unit boundary, investigate a new recharge project, or other choices by the unit operator (and unit committee) would be based on a dynamic relationship of aquifer data and responses.

4.4.7 Unit Boundary

The unit boundary is determined by the geologic conditions of the aquifer. The unit boundary is a three-dimensional (3D) shape, using depth and surface area components. For depth, the boundary would be determined by the strata making up a distinct water-bearing zone, and the discharge areas (like springs and stream baseflow). The boundary isolates an aquifer from other distinct water sources and other aquifers. For unconfined aquifers, the unsaturated and saturated portions of an aquifer would be included in the boundary. For confined aquifers, the boundary would include the region connected by piezometric pressure. Each isolated aquifer would represent a distinct unit and have a separate unitization agreement. Often, natural recharge zones are discussed as a vital component of the aquifer system.

However, the recharge zone is difficult to determine, making it a poor method of determining the unit boundary (Bredehoeft, 1997, 2007). Instead, the unit boundary is decided by the discharge locations, like springs, river baseflow, and wells. The natural recharge and recharge rate changes with variations in aquifer conditions. These are difficult to scientifically predict and verify. Recharge zones and rates are a potential source of conflict and might be an avoidable issue.

Additionally, surface water could potentially increase the size of the relevant boundary for the unit. Surface water can provide a significant source of recharge, either by the capture of streamflow or irrigation seepage (sometimes called "inefficiency"). Efforts to increase efficient irrigation of surface water, to improve stream flows, can have the opposite effect downstream or later in the season by reducing aquifer recharge and subsequent return flows into rivers (Kendy & Bredehoeft, 2006). For this reason, holders of surface water rights could potentially hold a significant number of artificial recharge shares in the unit. The planning stages of the CAGA would determine if including surface waters in the unit boundary would be beneficial to the unit's goals.

4.5 Unitized Substances and Shares

The substances that could be potentially unitized include any transresource of value in the aquifer. The primary substance supporting the formation of the unit would be water. Since water comes in various forms, qualities, and colors (blue water, green water, greywater, etc.; see Jarvis, 2014), each of these could be treated as separate interconnected shares. Surface water, groundwater, pore spaces, groundwater contamination, temperature, biological effects, subsidence, surface water interactions, waste, salinity, or any other transresources could potentially form share systems in a unitized aquifer. In the oil and gas context, oil shares are often separately determined from gas shares. For aquifers, the relevant transresources could be selected for conversion into shares. One aquifer may be suffering from depletion and contamination but have no real value in temperature (geothermal or groundwater cooling). The transresources selected for conversion into equity interests would be the relevant or necessary ones for accomplishing the unit goals, reflect the conditions in the aquifer, and represent how those transresources interact.

The determination of shares in an aquifer may be the most important and contentious negotiations during the development of the aquifer unitization agreement. These interests convert the water and property rights held by the community into collective interests in the aquifer. The outcome of this negotiation will determine the winners and losers in the unit. The choice of method ranges from simple, effective, and cheap approaches to complicated, but more equitable, market-driven systems. Like in an oil and gas context, the choice reflects the outcome of the CAGA negotiation.

Different potential approaches could be used to divide up the transresources in the aquifer based on local conditions, the opportunities in the aquifer, the social needs of the community, and the motivations of the participants. The interests would be designed to further the goals of the unit. Depending on the goal, the interests could come in various forms. The interests could reflect different forms of quantification, balancing the costs and benefits of implementation.

The selection of the equity interest method must be carefully chosen to reflect a compromise between purposes, fairness, and physical reality. These interests form a kind of property right, which "address the externality directly and link individual incentives with social objectives for resource use" (Libecap, 2008, p. 381). These shares reflect the reassignment of property rights into collective shares by contact, as described by Lai et al. (2016). These interests allow the revision of the previous system of (ground)water rights to better encourage behavior beneficial to the collective members of the aquifer and internalize externalities. That collective benefit, decided collaboratively by the unit committee, is enshrined in the goals of the unit plan.

The equity interests would be available to any group holding rights (property, surface, instream, tribal, or groundwater rights) within the unit boundary, including individuals, nonprofit groups, environmental groups, or even endangered species that have allocated water rights (for example, instream rights). "The integration of 'nature' or 'ecosystem services' into a [CAGA] would be a simple matter of listing ... nongovernmental organizations represent [ing] any facet of 'nature' in the agreement" (Jarvis, 2014, p. 29, modified for clarity). The inclusion of many parties would allow for the most effective outcome of the unit. Because each group would play a role in the unit, their interests would be directly tied to aquifer governance by their equity interests. Therefore, the method of determining unit equity interests is a critical, if potentially conflictive, step in the pre-unit negotiation.

4.5.1 Groundwater-Focused Share Systems

Because of groundwater's importance to the aquifer, a potential CAGA may focus on groundwater quantity, quality, and interactions with surface water. In these share systems, the aquifer is viewed primarily as a groundwater system. The social importance of groundwater as a resource makes it a good candidate for the basis of a system of shares. Groundwater-focused options are listed in Table 4.1.

The two primary issues with assignment of shares are the costs and complexity of developing the system balanced against the accuracy and effectiveness of the system of shares. A cheap and simple solution could be the use of land area as a rough approximation of aquifer transresources, akin to the land-size conversion of water rights suggested by Anderson (1983) and discussed by Smith (1986). Total pore space volumes may be appropriate for determining shares of a groundwater bank, recharge potential, or waste storage. Here, pore space volumes refer to the recoverable or effective storage potential, not the actual total volumes. Not all groundwater may be recovered from an aquifer, and some remains trapped due to capillary forces (Alley, 2007). Groundwater rights could form the basis for volume-based shares reflecting current demands on the aquifer. The currently stored volume of recoverable groundwater could allow a CAGA to distribute groundwater in a fossil aquifer under a managed depletion rationalization scenario. The shares could be simply designed for the introduction of water markets and banking, like the unbundled surface water rights discussed by Young (2015) but require no legislative changes to the underlying water rights.

The most complex, but effective, share system could be designed based on the effects of individual wells on the aquifer: a mass balance approach. Groundwater rights could be separated into the various components of discharge from the system, and further refined by the type of aquifer. Anderson (1983), Smith (1986),

Table 4.1. *The potential methods of creating aquifer shares. These options focus on groundwater and aquifer storage, but could be based on any transresource, like heat, carbon-storage potential, biological factors, or others*

Share basis	Description	Benefits	Costs
Land area	Estimation of groundwater or pore spaces by surface ownership alone	Cheap, easy to implement	Low accuracy
Pore space volumes	Total storage volume, appropriate for groundwater banking	Directly incorporates the aquifer's main use	Doesn't reflect groundwater volumes
Conversion of water right rates and volumes	Water rights are converted into volumetric totals	Cheap, easy to implement	Doesn't reflect the externalities of water consumption
Currently stored groundwater volumes	Volume of recoverable groundwater	Represents the in situ groundwater available	Doesn't reflect pore space storage potential
Water markets and groundwater banking	Interests defined merely for markets and banking	Designed for the ease of water markets and banking	May not reflect externalities
Mass balance conversion of water rights	Conversion of water rights into the components of the mass balance equation	Represents the physical effects of pumping	Expensive to determine, potential for conflict
Market tradable mass balance conversions of water rights	Market-based version of converted water rights into components of the mass balance equation	A marketable version, reflecting the physical effects of a water market and banking	Difficult to implement, expensive, potential for conflict

and Ghosh and Willett (2012) discuss separating water rights into replenishment and storage components, but this suggestion does not fully represent all the types of negative externalities of groundwater consumption that they suggest. Young (2015) separates surface water rights into a perpetual entitlement to an allocation and an annual actual allocation of water rights, without regard to physical negative externalities created by the water market.

A more accurate, environmentally sensitive approach follows the elements of the mass balance equation for aquifers, whereby each groundwater right would be separated into components of captured discharge: (1) stream capture, (2) storage depletion, (3) captured outflows, and (4) evapotranspiration. Often, recharge is used in groundwater budgets and suggested as a potential basis of groundwater property rights, but this value is difficult to determine and has little scientific validity (Bredehoeft, 1997, 2007). Instead, discharge is much easier to quantify and forms a stronger foundation for the issuance of shares. The elimination of natural recharge also provides a safety factor for the aquifer, allowing any recharge to provide a margin of safety in share allocations. Water marketing and banking can also be incorporated into the mass balance approach, allowing trade of storage depletion shares between well owners when state regulation prevents well use because of stream depletion and surface interference (using temporary transfers of the point of appropriation for groundwater rights) (Getches et al., 2015).

These shares are determined at the start of each irrigation season. For a given year, surface water right administration may result in groundwater right "calls" in conjunctive management states. Shares allocated to surface water depletion would not be available, leaving those users with remaining storage depletion shares available on the market. Likewise, when subsidence is a primary concern of a CAGA, the unit operator would limit the number of storage depletion shares it issues to aquifer users every year to manage, allowing users to purchase artificial recharge shares from other aquifer users to rebalance their own needs. The transfer of shares could incorporate actuarial science approaches to moving shares between wells instead of "injury" (Koda, 2007). The temporary transfer of water rights with the state agencies would be the duty of the unit operator managing the aquifer. The mass balance approach with marketing directly internalizes the physical externalities of a potential groundwater market and banking system, unlike the shares discussed by Young (2015).

4.5.2 Other Methods

The forms of equity interests created are not limited to the choices listed earlier and summarized in Table 4.1. The foundation of the unitization approach requires that the equity interests are scientifically based and verifiable. The examples listed earlier reflect a spectrum of choices, from simple and cheap to complex and rigorous. The more complex the method, the more equitable the resulting distribution of equity interests. Complicated calculation methods also produce more opportunity for conflict (English, 1996). Because no aquifer has been unitized, it is difficult to predict the result of the negotiation of hundreds of water right holders in a potential aquifer unit. Any method of determining equity interests will be a balance of the costs to develop against the benefits the system would create for the unit. These shares are a key component of the unitization approach and should be carefully considered in the negotiation of goals, economics, and politics of the aquifer community. Potential alternative methods could include other transresource aspects of aquifers.

Aquifers have the potential to act as purification systems using in situ bioremediation (EPA, 2013). Pollutants in the aquifer could also be more formally unitized, giving each participant a share of the contamination and the costs associated with it when assessing liability at Superfund sites. The Comprehensive Environmental Response, Compensation, and Liability Act (CERCLA) imposes joint and several liability on polluters that each contribute to the contamination of a single site (Oswald, 1995). Joint and several liability is often used when harms cannot be easily divided among contaminators (Oswald, 1995). Similarly, the contamination shares could be partially based on the outcomes of each party's contribution under CERCLA's joint and several liability.

Geothermal heat could be used as the basis for shares, unitizing the thermal properties of the aquifer (Jarvis, 2011; van der Gun & Custodio, 2017). Van der Gun and Custodio (2017) describe many different resources that all interact with aquifers, including oil and gas production, mineral extraction, and use of brackish groundwater. All of these examples share a common foundation: the in situ conditions of the aquifer, independent of the system of rights and laws determining access. The contractual equity interests are scientifically verifiable and selected based on the goals and purposes of the unit plan.

4.6 Collective Aquifer Governance Agreements, Systems, Laws, and Applications

Collective aquifer governance agreements will take the form of a negotiated, voluntary contract between the members of an aquifer community, which could include the holders of water rights (surface and groundwater), property rights, instream rights, environmental organizations, states, cities, counties, tribes, or any other relevant party that holds an interest in an aquifer. These parties make up the aquifer community, akin to the communities described by Shah (2009). A CAGA does not replace the existing system or rights, laws, or regulations, but seeks to coordinate their application and effects in an aquifer community. No contract may supplant existing law, but these agreements may be used to create new rights and obligations to achieve sustainable, holistic, and economically efficient resource use patterns (Jarvis, 2011; Lai et al., 2016; Max-Neef, 2005). Therefore, the contractual approach does not threaten to undermine or replace the current governance system, but can provide new opportunities, seeing the aquifer as "basket of benefits" rather than a wicked problem (Lawrence, 2010; Wolf, 2007). Collective aquifer governance agreements have the potential to move past the regulation-controversy feedback and reorganize the relationship between the social-legal and transresource systems in a new, effective way.

5

Determination and Redetermination

In the water business, we deal with deep dark holes, so you never
actually know what your sources hold.

—*Summit County, Utah, Water System Operator*
quoted in S. Moffitt, 2012

"Black swans" are rare events with extreme impact, followed by retrospective
predictability. *The Black Swan: The Impact of the Highly Improbable*, by Nassim
Nicholas Taleb, published in 2007, brought the concept to the popular literature.
Good examples of black swans are global economic crises that – after the fact –
"experts" claim were predictable ahead of time, based on a variety of indicators.
Looking back on the economic collapse, many "experts" indicated the financial
world should have seen the bubble bursting, based on the unsustainable market
fueled by the mortgage markets.

When it comes to the development of transresources like oil and water,
"wasteful haste" and "undesirable results" are "black swans" (Table 1.4). In the
1920s, the oil industry recognized wasteful haste as conserving the oil and gas
reservoir properties that were needed to bring oil out of the ground before the
advent of artificial lift. As oil fields were increasingly developed, wasteful haste
also included economic waste through the duplication of development and
operation outlay, many unnecessary wells, and a volume of production sometimes
unconnected to market demands (Mid-Continent Oil and Gas Association, 1930).

In the mid-1960s, Ostrom (1965) documented wasting groundwater storage
through seawater intrusion as an undesirable result in southern California.
Damaged and lost aquifer storage due to groundwater pumping eventually led to
widespread drying of wells tapping from compressible fractured rocks in eastern
Utah in the early 2000s (Jarvis, 2011). The Sustainable Groundwater Management
Act (SGMA) of 2014 identified other undesirable results as part of addressing
groundwater depletion in California (Table 1.4).

To combat waste and undesirable results with respect to conserving valuable subsurface storage, the oil and gas industry recognized that the hydrocarbons and storage were nonrenewable resources and incorporated a process to "determine" the volume of hydrocarbons that would be shared by tract participants, followed by "redetermining" the volume, rate of production, and boundaries of the reservoir.

Unfortunately, other resources stored in the subsurface, including groundwater and geothermal energy, were developed without much regard for whether the resources were nonrenewable – until, in some cases, either the resources were depleted, the storage properties of the aquifer were damaged, or both. Determination of the water availability is highly variable and relies on a broad spectrum of methods, such as estimating through 3D estimates of pore space storage, balancing the "safe" or "sustainable" yields through estimating water budgets and annual recharge, and caps on rates of production through surface acreage–based ownership; boundaries of the resource were often assessed and allocated after undesirable results occurred. Redetermination of the availability of groundwater and geothermal resources and aquifer storage are only reassessed and reallocated when undesirable results become problematic within depleting groundwater compartments identified by either continual declines in water levels or spring flows within index or observation wells and springs, within basins where groundwater and surface water are known or suspected on hydraulic integration, or within the political boundaries of conservation districts.

This chapter compares and contrasts the determination and redetermination processes used in the oil and gas industry versus groundwater development. It will be shown that waste does not regularly occur when developing oil and gas reservoirs because of the determination and redetermination programs associated with unitization. Conversely, undesirable results are commonplace where groundwater is developed because of a lack of regular redetermination of groundwater and aquifers.

5.1 Initial Determination

Sato and Crocker (1977) summarize four models currently applied to determining interests or allocations in subsurface resources such as oil, gas, geothermal energy, and groundwater including the rule of capture, the reasonable-use rule, the correlative rights doctrine, and the appropriation system. Regardless of the transresource that has been discovered through drilling, the first step is to determine the size of the discovery and allocate interests in the resource.

In the oil and gas sector, the shares of reserves are allocated to each of the licensed blocks or "tract participants," which are maintained until redetermination or are adjusted by signature of the unitization agreement. These "fixed" shares are negotiated rather than decided by relying upon a technically determined figure based on the best available technical knowledge at the time of drilling. There are no industry standards for determination – the method depends on the type of hydrocarbons in place and the quality and type (porous media versus fractured rock) of the reservoir. The general approaches include estimating the hydrocarbon pore volume through geophysical log interpretation, downhole formation testing, or laboratory measurements; hydrocarbons initially in place; moveable hydro-carbons initially in place; initial recoverable reserves; and economically recoverable reserves. The oil and gas industry prefers opting for more straightforward methods. The initial well testing data can be held "tight" or confidential by the well owner, but are typically made available to the public through a state or provincial agency after the "tight hole" status negotiated between the regulatory agency and well operator expires.

Unless there are agreements in place before discovery, fixed interests typify the outset of development. For example, water laws and state regulations govern water rights and the development of groundwater and aquifers following an initial flow or pump testing of a new well. These rights, rates, and duties associated with groundwater development are typically fixed interests for all time, and are subject to abandonment in some jurisdictions if the water right is not "used" to its fullest extent. As a consequence, the initial testing data are usually publicly available on well logs or well completion reports for drinking-water wells filed with a state or provincial agency.

The success or failure of the fixed-interest approach to transresource governance depends on the confidence in the technical parameters without benefit of drilling development wells, which may require something of a "leap of faith" (Asmus & Weaver, 2006).

5.2 Redetermination in Oil and Gas Reservoirs

Unitization of oil and gas resources recognized early on that "change happens," and that responding to changed subsurface conditions required a regular redetermination of interests, as conceptualization of the reservoir geology became refined and as more subsurface data were collected. In the case of smaller reservoirs, pooling of the smaller reservoirs can lead to developing the smaller separate fields as one reservoir.

One of the best summaries of the oil and gas approach to dealing with change is by groundwater hydrologist John Bredehoeft (2005, p. 44):

Petroleum engineers have perfected the art of short-term reservoir predictions. They look at the models from a pragmatic perspective. They create the mathematical model from their best understanding of the prevailing theory. They then apply the model to a particular petroleum reservoir. They adjust the parameters to match an observed history of reservoir performance – they call this match a "history match" (rather than model calibration). They then use the history-matched model to make a prediction of reservoir performance. However, they have caveats regarding their predictions. The rule of thumb is not to rely on the predictions much beyond a period equal to the period of history match. In other words, if one matches a 10-year reservoir history the engineer has some confidence in making a 10-year prediction. Beyond the 10-year prediction the engineer questions his confidence in the prediction. Petroleum reservoir engineers avoid making claims that they have the correct conceptual model. They say simply we did the best we could to create what we think is an appropriate model of the reservoir. We will use this model to make a prediction of performance in which we have confidence, for a period equal to our history match. These rules of thumb could well be applied to groundwater analyses.

Asmus and Weaver (2006) indicate that if the unitizing parties are comfortable with a particular basis for determining tract interests, then redeterminations of tract interests will be completed on the same basis. Parties may agree to set initial tract interests such as acre-feet of reservoir rock. Redeterminations can be based on a more precise basis once additional data are available, perhaps focusing on recoverable reserves, which are different from stored reserves estimated based strictly on volume of reservoir rock.

Two approaches to redeterminations in the oil and gas sector are commonly used that describes technical rules, starting with common and agreed databases and steps to be used in calculating tract participation (Asmus & Weaver, 2006). According to Worthington (2018), the most common bases are initial hydrocarbon pore volume (IHPV), at reservoir temperature and pressure, and hydrocarbons initially in place (HIIP), at surface conditions. Together these static bases account for about 85 percent of unitizations outside North America. Minority options are estimated recovery (ER) and estimated economic recovery (EER) of hydrocarbons during field life.

A second approach develops general guidelines to establishing ground rules, which allows greater freedom to work out the technical details at a later date. Both approaches include data cut-off points where no further data are collected for inclusion in the redetermination. If an agreement on the redetermination process cannot be reached, the decision can be left to a redetermination expert, the oil and gas field operating committee, or the decision of the majority of tract interests.

The timing of redeterminations for oil and gas reservoirs typically requires five or more years of production history. Later redeterminations are negotiated and can be based on a stated period of production history, or as a result of the acquisition of other new data, such as data from the drilling of additional wells or geophysical surveys. Asmus and Weaver (2006, p. 86) provide a summary of international redetermination case studies indicating there are no hard and fast rules regarding timing and data acquisition:

- First redetermination upon the earlier of final development well or commencement of commercial production; second redetermination after two years of production.
- Redetermination to be conducted on or before third anniversary of government approval of the unitization.
- First and second redeterminations to be conducted on specified dates. No field included unless it has at least one year's production history.
- Redetermination requires "relevant new technical information."
- First redetermination on a specified date; second redetermination six months after the drilling of the last unit well.
- Redetermination to be conducted upon achieving a specified cumulative production level and completing one new well.
- Redetermination process began upon execution of the agreement.
- First determination after the drilling of six development wells; second redetermination upon request between two specified dates.
- Redetermination to be conducted upon the earlier of five years from the issuance of the license or completion of 90 percent of the development drilling program.
- Redetermination upon request, but no more than one every five years.

There are no limitations to the number of redeterminations; however, redeterminations involve extensive review of technical data and expert assistance, and expensive and time-consuming litigation or arbitration may result where the parties do not agree upon the adjustments. For example, the redetermination of Prudhoe Bay in Alaska required over five years and tens of millions of dollars (Asmus & Weaver, 2006).

Redetermination leads to adjustments to the unit area. Two types of boundaries are estimated at the time the agreement is entered into between the tract participants of the unit area (Jarvis, 2011, p. 625):

- Voluntary units: agreements among interested parties that can be undertaken for exploration or conservation; and
- Geographic units: typically applied where poor well control precludes defining boundaries of productivity, or in areas of complex geology.

Redeterminations can, and usually do, change the boundaries of the unit area because more subsurface data become available (Association of International Petroleum Negotiators, 2006). The types of unit boundaries resulting from redeterminations include

- compulsory or conservation units: a high level of knowledge of physical characteristics of the "pool" is required for conservation;
- geologic units: typically applied based on geology, productive area, lease position, precedent in a field, producing horizon or trend or economics (Jarvis, 2011, p. 625).

Asmus and Weaver (2006, p. 84) also indicate that "redeterminations merely reallocate existing value rather than create new value. In practice, only one or two redeterminations are permitted."

Given the technical nature of oil and gas reservoirs, coupled with the financial implications of redeterminations, the redetermination procedures are typically prepared by technical experts rather than by lawyers. Lawyers review the technical redetermination procedures from a legal perspective. As a consequence, lawyers need to sit down with experts to ensure they understand the intent and content of redetermination procedures. As we shall see in later sections, this same transdisciplinary approach was used for the development of the 2008 Draft Articles of Transboundary Aquifers, as described by Eckstein (2017).

5.3 Redetermination for Groundwater and Aquifers

In contrast to the structured process of determination and redetermination employed by the oil and gas industry, there are no "cookbook" approaches to ascertain whether or not "undesired results" develop as a result of groundwater development and aquifer storage depletion. The problem rest with the complexity of how groundwater is stored in an aquifer; hydrogeologic and political boundary conditions; and politics in terms of desired future results. Water-level monitoring, production monitoring, water quality monitoring, recharge assessments, water rights maintenance, property rights assurances, and reallocating through pro rata sharing all factor into redetermination processes.

5.3.1 Direct Methods

5.3.1.1 Historic Pumping Records, Southern California

One of the earliest examples of a unitization-style redetermination process for groundwater and aquifers is the germinal work on the groundwater commons in southern California by Ostrom (1965). This exercise in public entrepreneurship involved (1) the organization of a private association to provide a forum for the discussion of joint problems; (2) the formation of a municipal water district to furnish an alternate supply of surface water; (3) the initiation of a limited form of pro rata rationing of groundwater supplies; (4) the determination that a freshwater barrier to prevent saltwater intrusion was physically feasible; and (5) the creation of temporary institutional arrangements to operate a partial barrier against the sea.

One of the many agreements between the groundwater users included the addition of a provision to enable a party to pump their relative prescriptive right for 1961 during every fifth year in order to protect claims to groundwater rights developed prior to the signing of an agreement while a dispute was litigated in one of the groundwater basins. In this situation, courts determined the relative rights on the basis of past production records. Parties would be allowed to request a redetermination of the production histories upon which their relative rights were based. This gives a signatory party the right to challenge the validation procedure used by an expert and the opportunity of a new determination made by the basin watermaster.

5.3.1.2 Concurrency, Summit County, Utah

Jurisdictions across the United States are crafting policies that specifically require "proving" water availability for housing developments (California, Colorado, Texas, Utah) and new agricultural uses (California). Some counties are also weighing interference between proposed developments and senior surface water rights through uncontrolled pumping of groundwater through domestic wells (Washington). Elsewhere, counties are asked by state governments to develop groundwater management plans to ascertain the availability for other high value uses such as permitting short-term sales of groundwater appropriated for agricultural uses to the drilling industry for hydraulic fracturing or "hydrofracking" (Wyoming).

The fragmented nature of water and land use laws at the level of individual counties, states, provinces, and countries is leading to a new paradigm in water planning and management that focuses on a bottom-up approach instead of the traditional top-down approach. Concurrency laws are one of the most effective tools for linking water availability and land use at local jurisdictional scales. Since the 1970s, concurrency laws in the United States have typically focused on the availability of public facilities, such as schools, roads, sewers, and water supplies, to accommodate rapid growth.

The nexus between the planning goals and groundwater resources looks good on paper, but like water rights, is only as good as the paper it has been written on. Under ideal conditions, municipalities and counties would have known the sustainability and carrying capacity of the groundwater resource when comprehensive plans were originally being formulated. Lot sizes and densities could have been set at a level that would not result in mining of the groundwater resources. Land use limitations could have been designated in areas where the hydrogeology indicated that the groundwater resources were vulnerable to contamination. However, acquisition of these data is the "unreachable star," as the quantity of information to make these assessments is generally not available. Municipalities and counties are faced with difficult choices. They can neglect or ignore potential problems until local users demand a solution, as observed in Wyoming and Utah where developers are responsible for supplying water to their developments, but only do so under duress. Or the municipalities and counties can require applicants for new land uses to demonstrate that their proposed developments will not result in harm to the groundwater resource as outlined by the concurrency requirements in other states described by Strachan (2001).

Concurrency laws have evolved to focus more on exurban growth and groundwater availability. The change came when the requirements shifted the linkage of land use from groundwater storage to groundwater recoverability. The change in this instrument came about due to highly variable well yields unrelated to groundwater recharge or depletion; rather, the well yields were related to damaged and lost aquifer storage.

Implementation of concurrency ordinances requires making decisions in the face of uncertain data. While periodic retesting for redetermination of groundwater availability may reduce risky water relationships, it risks antitrust restraints. However, the only methods to ascertain whether or not groundwater and aquifer storage depletion are occurring include production and water-quality monitoring. In some locations such as Utah and, until recently, California, well and groundwater data may be considered proprietary. Likewise, the efforts at regionalizing groundwater-based systems that do not have sufficient water supplies to support continued growth have been considered as a means of monopolizing the water market, thus triggering concerns and court cases focusing on violating antitrust laws.

Carlton (2007) describes a case study in Utah in which the authors were directly involved, so the following discussion fills in some gaps for the legal case study. Land developers were required to secure a commitment from a water company to acquire building permits from Summit County. Water systems dependent on wells and springs tapped the fractured rock aquifers located in a complexly deformed geologic province known as the Overthrust Belt. Hydrogeologic analyses of well yields revealed groundwater compartments reflected by seasonal changes in pumping rates, changes in water quality that eventually made some of the well water unsuitable for a drinking-water supply without either treatment or blending, and water-level changes consistent with groundwater depletion. Investigations of groundwater recharge and residence time using noble gases, tritium, and terrigenic helium by Klein (2008) determined that groundwater developed by some water companies ranged in age from six to fifty years, with an average age of over thirty years, supplementing the notion that groundwater mining could occur.

Water companies providing the "water letters" were required to prove they had the water, after undergoing a concurrency rating. The concurrency rating system was designed to build upon the initial testing of a new well or developed spring using standard water-well industry practices, in the cases of wells using controlled pumping tests as described by Driscoll (1986). The concurrency rating was based on a "what you see is what you get" approach with a small conservation reserve applied to the rating as one means to preclude groundwater depletion. This approach permitted the identification of problems with pumping equipment and appurtenances, changes in well efficiency due to chemical or microbiological encrustation, and changes in the aquifer storage properties commonly associated with fractured rock aquifers leading to a 50 percent reduction in well yield after just a few years of installing a new well (Jarvis, 2011).

A private water company in the Snyderville Basin of Summit County filed a lawsuit against a regional water system, with the concurrency or "redetermination" expert, among others associated with the development and implementation of a concurrency ordinance, suggesting that the process "restrained trade and monopolize the private water market," and that the Utah Antitrust Act did not refer to either a county or a special district, nor were the defendants' actions authorized or directed by state law (Carlton, 2007, p. 689).

The lawsuit can be only be described as an odyssey, given that the fortunes of both parties changed frequently. The lawsuit was settled by mediation in 2011, with the water company initiating the lawsuit attempting to drill one of the deepest water wells in Summit County to meet the deficiency in water supplies for their commitments; however, the yield from the deeper aquifer fell short of their required water needs.

While the concurrency ordinance was under consideration for repeal at one point in time, the county elected to continue with the periodic redetermination process for private and public drinking water systems, albeit with minor changes in the testing frequency and more transparency in reporting of the concurrency reports. Neighboring water systems dependent on wells tapping the fractured rock aquifers continued to deplete groundwater compartments. The systems eventually connected to water systems with water surpluses through supply augmentation, essentially creating a smaller number of larger regional water system "units." The ironic result is that the water system that initiated the initial lawsuit and drilled the deep well eventually connected to a larger regional water system in order to lease water from nearby water systems with surpluses (Brough, 2019). The result was the de facto creation of regional drinking water "units" within Summit County, Utah.

5.3.1.3 Redetermination through Incentivized Managed Recharge, Eastern Snake River Plain, Idaho

Basalt aquifers are most productive aquifers in the Snake River Plain of Idaho, and they underlie most of the 10,800-square-mile eastern plain; the basalt aquifers are also linked to springs that contribute to flows in the Snake River (Lindholm, 1996). According to Weiser (2018), the aquifer had been dropping at an average rate of about 214,000 acre-feet annually since 1952. Various water rights holders settled a lawsuit in 2015 requiring the state to begin replenishing the groundwater utilized by farms, cities, and fish hatcheries.

Tuthill and Carlson (2018) describe an innovative process of incentivized managed aquifer recharge (IMAR), utilizing ownership of marketable ARUs in the Eastern Snake Plain Aquifer in Idaho. The IMAR concept is simple: ARUs are comparable to contracting for space within a federally constructed reservoir. The space holder is guaranteed space, not water. An ARU is filled annually with a managed aquifer recharge (MAR) event, if surface water is available for recharging the aquifer.

Aquifer recharge units are described as "fungible, have value, can be bought and sold and are directly analogous to the space acquired in a surface reservoir" (Tuthill & Carlson, 2018, p. 11). An ARU is aquifer storage space "redetermined" directly through MAR volumes that are measured during the artificial recharge process; the measured volumes are allocated to ARUs which represent one acre-foot of virtual space in the aquifer. The evacuation of ARU storage generally relates to a pumped volume attributed to the ARU holder. And, like oil and gas units, an organization that is owned and operated by ARU owners is established under state law to accomplish long-term management of the ARUs and associated MAR within the groundwater basin. One example described by Tuthill and Carlson (2018) focuses on American Indian Reservations within the western United States. Much of this land is located in headwater areas, where opportunities exist for MAR.

5.3.2 Indirect Methods

5.3.2.1 Three-Dimensional Management in a Texas Groundwater Conservation District

Texas is famous as an example of the "rule of capture" in groundwater law, where landowners could pump as much water as they wanted without waste, under surface-based correlative rights; but as with many states, the definition of "waste" is poorly constrained. This led to problems for landowners where geology got in the way, with aquifers having variable thicknesses across regions. Landowners located on thick sections of saturation of an aquifer could capture more groundwater than landowners overlying thinner sections of the same saturated aquifer. The result was some properties had more "rights" than others, leading to surface owners overlying the thickest parts of aquifers "selling" their rights to municipal suppliers. This could eventually cause problems for groundwater conservation districts by meeting or exceeding the caps on production designed to conserve groundwater. According to the State of Texas Water Code Ann. § 36.0015,

Groundwater conservation districts created as provided by this chapter are the state's preferred method of groundwater management in order to protect property rights, balance the conservation and development of groundwater to meet the needs of this state, and use the best available science in the conservation and development of groundwater through rules developed, adopted, and promulgated by a district in accordance with the provisions of this chapter.

According to the Texas Water Development Board, the Carrizo Aquifer is an important aquifer supplying water to Guadalupe County Groundwater Conservation District. The Carrizo Aquifer is primarily composed of sand, locally interbedded with gravel, silt, clay, and brown coal. The aquifer can reach upwards of 3,000 feet in thickness, yet the freshwater saturated thickness of the sands averages 670 feet. Blumberg and Collins (2016) posited that the landowners who owned property overlying the thick saturated sections of the Carrizo Aquifer essentially controlled the aquifer and precluded competition from other ground-water rights holders who owned property with thinner saturated sections of the aquifer. Recognizing that geology was getting in the way of equitable and reason-able groundwater management, the Guadalupe County Groundwater Conservation District developed a 3D groundwater volumetric rights allocation system by "mapping the resource and equitably dividing it based not on flat surface acreage but rather on the available volume of saturated Carrizo Aquifer sand under each tract" (Blumberg & Collins, 2016, p. 70). Under this new management model, owners of thin saturated sections of the aquifer also have rights that permit participation in the marketplace.

The 3D management system and redetermination processes were developed using a GIS database and 3D model of the saturated thickness. These datasets were created using contour data, water-level measurements and other hydrogeologic data, coupled with a digital property surface map, and were ultimately used to assign a percentage of the entire saturated section volume to the properties overlying the Carrizo Aquifer. The conceptual model Blumberg and Collins (2016, p. 75) describe is simple:

[T]he aquifer may be thought of as a static-level lake with a certain inflow (recharge) that is divided fairly to all property owners "on the bank of the lake." The constant-level lake is owned by no one, only the inflow (recharge). The inflow is distributed pro rata, depending on how many feet of bank each owner owns and the "lake" (i.e. the aquifer) is only a temporary holding tank for the inflow.

They acknowledge that the management model is for idealized hydrologic systems and does not account for artesian pressure in the aquifer. They also note that for complex aquifers with confining units and other parameters leading to anisotropic and heterogeneous conditions, the groundwater volume models must also account for hydraulic conductivity.

Blumberg and Collins (2016) reported that the Guadalupe County Groundwater Conservation District redetermined that the total annual allowed production should equal 62.5 percent of the assumed annual recharge, yielding a maximum annual extraction volume of 12,600 acre-feet. While the redetermination process could yield firm production rates, they also note that this desired rate also can be modified due to local politics and desired future conditions.

Blumberg and Collins (2016) acknowledge that the Guadalupe County Groundwater Conservation District's "redetermination" of allocations under the new management model closely coincides with the precepts of "pooling" common oil and gas leases. The computer-generated redetermination process was built upon a defined pool of rights that applied to all water owners, with the goal of reducing the risk that a groundwater conservation district, or "aquifer unit operator," would favor one set of aquifer users over another. Blumberg and Collins (2016) indicate that by defining a pool of water volumes, "aquifer owners" can market groundwater on their own without burdening the regulatory agencies and groundwater conservation districts.

5.3.2.2 Redistribute Reductions in Escalante Valley, Utah

While the notion of experimenting with unitization in western Utah was dismissed by some in the region as too utopian, as described by Clyde (2011), in 2010 farmers and ranchers in the Escalante Valley of southern Utah faced water-level declines over 100 feet that occurred over the past 50 years. Lost aquifer storage was revealed through subsidence: earth fissures tens of feet in length and up to six to seven feet in depth were found in the Escalante Valley (Hansen, 2010). According to the State of Utah (2012), the safe yield for the unconsolidated and semi-consolidated sedimentary deposits composing the groundwater basin was determined to be approximately 34,000 acre-feet per year; the current average depletion is estimated at approximately 65,000 acre-feet per year.

In contrast to the western Utah situation, legislation passed in Senate Bill 20 in 2006 granted the state the authority to manage the groundwater where paper water rights could not be met by wet water production by transferring the resource management to a local water management district. The Escalante Valley became the initial test case for changing groundwater use (Hansen, 2010). While the State of Utah introduced a plan to reduce groundwater use by 50 percent over a period of ninety years, the water users found the plan unacceptable and "pooled" their water rights to share in reductions in water use by voluntarily forming a "unit" – the Escalante Valley Water Users Association – to manage withdrawals based on a system other than priority dates. Their plan was simple: redistribute reductions across irrigation uses by 5 percent every twenty years with the goal of achieving safe yield of the Escalante Basin by 2190 (Hansen, 2010; Keiter & Ruple, 2011).

The perception that indirect measurement of hydrologic and aquifer storage changes would take nearly two hundred years to prove out was met with skepticism by many parties. Keiter and Ruple (2011) indicate the agreement was unclear as to which parties would shoulder much of the burden of implementing the agreement. For example, if senior water rights holder ceased water "use," would that constitute abandonment and forfeiture under Utah Water Code? Did a voluntary agreement have any power for enforcement? Could a party withdraw from a volunteer agreement? What value was there in entering into a voluntary agreement by a water right holder with a senior priority?

With the Escalante Valley Water Users Association unable to formalize a workable voluntary agreement, the state engineer developed a groundwater management plan for the Enterprise Area. Keiter and Ruple (2011) describe the plan as an apparent compromise between the timeline advocated by the Escalante Valley Water Users Association and earlier timelines proposed by the state engineer. The plan calls for two phases: Phase I requires consumptive reductions of 5 percent over each of two 20-year periods; phase II requires comparable 5 percent reductions but over seven 10-year periods, concluding with a 3 percent reduction over the final decade. Under the plan, as adopted in 2012, safe yield would not be achieved for 120 years. Reductions would be implemented based on water-right priority, effectively requiring junior water users to either forego withdrawals or acquire water rights from senior water users.

5.3.2.3 *Redetermination of Sustainable Yield, California Sustainable Groundwater Management Act*

Before the passage of SGMA in 2014, groundwater pumping regulations were essentially nonexistent. According to the California Department of Water Resources, groundwater comprises approximately 38 percent of California's total water supply, in drought years groundwater contributes up to 46 percent of the state's annual water supply or approximately fifteen million acre-feet per year (Rudestam & Langridge, 2014). Undesirable results from the unbridled pumping remain manifold, with perhaps the most alarming being ground subsidence, seawater intrusion, and deeper well drilling due to chronic lowering of water levels, among others. These led to a prioritization of groundwater basins requiring some form of "sustainability plan" developed by a groundwater sustainability agency that included a "water budget" and "sustainability goals" looking forward fifty years.

But what constitutes "sustainability" and "sustainable yield," and how are these planning metrics calculated? Does a single number serve as a pumping goal? Will the pumping goals change with time? These questions were investigated in the early stages of implementing SGMA by Rudestam and Langridge (2014), and the answer to the question was "what do you want it to mean"? Regardless of what side of the table a hydrogeologist is sitting on when it comes to debating "what it all means," they will all agree that calculating inflows and outflows is complex, is not an exact science, and typically requires expensive field monitoring to develop extensive and costly hydrogeologic conceptual models to transform into numerical modeling studies in virtually all basins. And as one hydrogeologist interviewed by Rudestam and Langridge (2014, p. 95) opined, "if you have one well for an entire distribution system that is located on the coast, the sustainable yield would be much lower to prevent salt water intrusion than a system that had multiple wells farther from the coast," and that sustainable yield "needs to be dynamic, flexible, and case specific to accommodate variations in timing and pumping."

Redetermination through revision and refinements in models is a regular occurrence during the implementation of a groundwater sustainability plan (California Division of Water Resources, 2016). The scheduling of a redetermination process typically focuses on when models are updated and refined and varies as new data are made available, but the recommendations are at least made annually and during the five-year review process.

5.4 Conclusions for Determination and Redetermination

Acknowledging the complexity and cost of redeterminations, Asmus and Weaver (2006) indicate that the oil and gas industry is foregoing redetermination clauses in unitization agreements in areas where good seismic and well-control data are available or where the field is of marginal commerciality that does not justify the cost of any redetermination. However, they also indicate that proceeding without redetermination clauses presents substantial risks for each tract in a situation where few data are available at the time the unitization agreement is negotiated. Worthington (2018, p. 6) echoes this philosophy "despite the mitigation of wastage and the assurance of greater fairness and equitability," yet "where the prize is sufficiently large, unitization and equity redetermination of a straddling petroleum accumulation will always prevail."

For surface water systems serving agriculturalists, Young (2014, p. 32) tacitly acknowledges the use of redetermination principles through the periodic revision of water sharing plans where "individual adjustment to changing circumstances is facilitated through trade in entitlements and allocations." Clyde (2011, p. 10) suggested that the unitization process could also work in urbanized areas where municipalities compete for a common water supply. He also indicates that at the smaller levels of competing retail water systems that use the same groundwater resource and aquifer, "the unitization approach may well save the day."

Clearly, one approach to determination and redetermination of aquifers does not fit all situations due to the hydrogeologic complexity, variable storage characteristics, and politics.

New instruments of determination and redetermination for groundwater and aquifers must integrate groundwater and aquifer storage. Concurrency instruments are one direct method of determination and redetermination that hybridizes land use, groundwater use, and aquifer storage. Variations of direct methods can also link past rights and use with current use.

Unitization instruments are more robust indirect and direct methods that hybridize the processes of exploration, storage, and extraction of transresources. These instruments look beyond just groundwater by linking past, current, and future uses of aquifers.

6

The Role of the Expert

A source of unclean information can poison many minds.

—*H. L. Doherty (1924)*

I know what's the matter with the oil business. The lawyers have gone
to practicing engineering, and the engineers have gone to practicing
law. It appears that a new profession has been created, the members of
which will be awarded the degree LG – the mark or rank of
the Lawgineer.

—*Unknown, in Hardwicke (1948, p. 65)*

Hardwicke (1948) wondered what kind of queer creature would be depicted if a
cartoonist should portray a typical Lawgineer or rather a bogus one, and received
the following suggestion (which we commissioned a cartoonist to draw, as depicted
in Figure 6.1):

[S]urely a Lawgineer would be a long cadaverous-looking fellow with an unmistakable
leer on this face; he would have a large, square head upon which would rest unsteadily a
box-shaped hat with the word "Constitution" written across it; for his arms, he would
have in flowing script the words "Immaterial" and "Irrelevant"; in his right hand he
would hold a bailer full of "Bottom-hole Pressures" while in his left he would hold a
bucket full of "Permeability"; he would have wobbly legs composed of the same flowing
script , forming the words "Aforesaid" and "Whereas"; and stuck through his belt, labeled
"Vested Rights," would be a slide rule called "The Rule of Capture," with which to
measure "Potentials."

Hardwicke (1948, p. 66)

It is clear the common problems of conservation and overproduction facing the
oil industry in the early development of oil and gas unitization were recognized
and the industry had the prescience of moving beyond a strictly disciplinary level
of understanding of laws by courts versus physical and economic laws to a
transdisciplinary paradigm. According to Max-Neef (2005), transdisciplinarity is

Figure 6.1 1948 Lawgineer

more than a new discipline; rather, it is a different manner of seeing the world that is more systemic and more holistic. Zhao and Anand (2013) also refer to these skills as "boundary spanning." Transdisciplinarity addresses several levels of reality and through several levels of organization. Max-Neef (2005) describes the different levels, ranging from "what exists," incorporating fields of study such as geology and economics, to "what we are capable of doing," that focus on the fields of engineering and commerce. The third level addresses "what we want to do" through planning, law, and policy. The top level refers to "what we must do" based on values, ethics, and philosophy. This is the foundation of the concept of "unitization" of oil fields that was developed to protect the "corresponding rights" or "sovereignty" of all pore-space owners in the unit and to not waste valuable pore space (Jarvis, 2011).

We envision the modern transdisciplinary Lawgineer focusing on groundwater or aquifer management as a physically fit person with a smiling face. She would have a rounded head on which would rest a baseball cap with the word "entrepreneur" on the patch. For her arms, she would have in bold font the words "repurpose" and "redetermination." Her right hand would hold a laptop computer with "aquifer model" on its screen, while in her left hand would be a rolled document titled "agreement." She would have strong legs composed of the same bold font forming the words "Collective" and "Governance." And stuck through her belt, labeled "water rights," would be a rolled map labeled "aquifer communities." Harkening back to her Texas roots, she wears a string tie, with strings labeled "public trust" and "public interest," as depicted in Figure 6.2.

Asmus and Weaver (2006) and the AIPN (2006) indicate that the redetermination expert essentially serves as a third party to assist with reaching agreements related to the redetermination process. The expert could be handed the entire redetermination process from the beginning, with the tract or aquifer participants abiding by their decision, but few participants are prepared to allow a third party complete control. Redetermination experts could be sole practitioners with transdisciplinary credentials, or a company with the resources and expertise to complete work.

Experts should be appointed as redetermination experts, not arbitrators. However, the distinction is not clear cut between offering an opinion, by which the parties agree to abide in order to prevent a dispute from occurring, and an arbitrator who has to decide on the respective merits of competing claims when a dispute has arisen. Unitization agreements such as those developed by AIPN (2006) may have procedures for the selection of the expert. Asmus and Weaver (2006) determined that most of the sample unitization agreements they examined from international settings contained redetermination clauses that referred redetermination disputes to an expert or experts. They also suggest it is best to select an expert early in the redetermination process before positions become entrenched.

Funding shortfalls, the uncertainty associated with the quantitative characteristics of groundwater and aquifer systems, and increased use of numeric groundwater models as necessary components for informed groundwater management decisions, yield a growing frustration with the dueling expert situation (Jarvis, 2014). Groundwater and aquifer data are considered proprietary information in some jurisdictions, thus leading to accusations of monopolies of aquifer resources when companies are forced to comply with redetermination requirements.

Figure 6.2 2020 Lawgineer

Do "lawgineer" redetermination experts exist? For redeterminations in the oil and gas sector, one of the most prominent experts and a transdisciplinarian focusing on unitization is Paul F. Worthington. We first were introduced to Worthington's formation evaluation work in the late 1970s and early 1980s with BP, where some of his publications focused on well-log analysis applied to groundwater. His degrees include a PhD in engineering geophysics and a higher doctorate (DEng) in geoengineering studies, both from the University of Birmingham, UK, and he later earned a master of laws (LLM) in Petroleum Unitization Law from the University of Reading. He now serves as a consultant, where his main interests are integrated technical and strategic studies for unitization and equity redetermination, reserves estimation, and the identification of hidden pay, as well as the petrophysics of problematic reservoirs.

We are not certain of comparable "lawgineers" focusing their practice exclusively on groundwater and aquifers, but we see this as an emerging field of specialization, as it becomes increasingly recognized that the utilization and repurposing of aquifers and aquifer storage spans many different disciplines.

7

The Next Transresource

The Emerging Wars over Pore Spaces

Thirty spokes share the wheel's hub;
It is the center hole that makes it useful.
Shape clay into a vessel;
It is the space within that makes it useful.
Cut doors and windows for a room;
It is the holes that make it useful.
Therefore profit comes from what is there;
Usefulness from what is not there.
(—Lao Tsu, quoted in M. A. Max-Neef (2005))

A theme of this book is the emphasis that the storage spaces of an aquifer may be just as important as the groundwater. While the debate over the public or private ownership of water continues, who owns the pore spaces? The question is critical for the application of unitization to aquifers, since the "unit" would be defined by the aquifer as a whole and not the presently occupied pore spaces filled with groundwater. Aquifer storage space, not groundwater, could become the new front in the water wars of the western United States and beyond.

Surface water conflicts have been exhaustively litigated, and few unexplored legal approaches remain. For groundwater and aquifers, inconsistencies between water and property law are just beginning to be explored. Subsidence and aquifer storage are examples of the emerging realization that water and property law are not distinct legal subjects. Because groundwater and the aquifer are inexorably intertwined, the effects of water law have direct impacts on the property-based rights to the solid portions of an aquifer.

7.1 A Brief Review of Science: Groundwater Hydrology

An important difference must be drawn between groundwater, aquifers, and pore spaces. Aquifers are often incorrectly associated with the water below the earth's surface. In truth, the aquifer is the portion of the subsurface *potentially capable* of producing usable quantities of groundwater. The aquifer is not the water but the container holding the water. In unconfined aquifers, the water table moves vertically as water is withdrawn or added. The portion of the aquifer above the water table is unsaturated but still capable of holding water. Unconfined aquifers are typically near the surface and may reach the land surface. In contrast, confined aquifers are separated from the surface and overlying unconfined aquifers by an impermeable sediment layer, like shales and clays. In confined aquifers, the pressure of the water counteracts the pressure of the overlying sediments. As water is withdrawn from a confined aquifer, the pressure within the pore spaces is reduced. A reduction in pressure within pore spaces, or fractures, can cause permanent compaction of the confining bed and aquifer, reflected at the ground surface as subsidence. An aquifer may be drained, refilled, depressurized, or permanently compressed (leading to land subsidence). When pore spaces, fractures, or conduits are filled, drained, compressed, or depressurized by water users, these effects, which have real world impacts, are not defined or discussed by a water right permit.

7.2 Unitization and the Pore Space Question

The question of aquifer pore space ownership is a key question for the application of unitization principles to groundwater governance. Private ownership generally includes the right of use and enjoyment, the right to transfer, and the right to exclude (Sprankling, 2008). All three of these components are related to development and use of aquifer storage spaces and unitization. Unitization agreements, as used in the oil and gas industry, engage all three of these components, by transferring the right to use the reservoir to the unit operator voluntarily (waiving the right to exclude).

As discussed in other portions of this book, unitization in the oil and gas industry treats the entire *reservoir* (including all brackish, gas, and otherwise occupied areas of the reservoir) – not the *oil* – as a single unit. The primary benefits of unitization stem from this fundamental shift in perspective. By manipulating the substances within the entire reservoir to increase efficiency, the net benefit to all participants in the unit increases. At this point in the development of unitization principles, the ownership of the storage spaces within a reservoir are nearly universally considered as a private good, even when filled with brackish *groundwater*.

For fresh(er) groundwater closer to the surface, the ownership of the storage spaces within an aquifer is still an open question. We could opine that the public ownership of the groundwater within aquifers has complicated the treatment of aquifers, and that shallower aquifers often have more connections to surface waters, where more public interest issues are presented. Unfortunately, we cannot provide a clear answer.

Despite the issues with identifying aquifer pore spaces as purely private, purely public, or a mixture of both, we are not convinced that private ownership of aquifers is a prerequisite for aquifer unitization. For example, if we suppose that aquifers are purely public spaces free for all to use as groundwater storage, the benefits of unitization still exist (maximizing the utility of the aquifer for the benefit of those who can access it). Being a purely public resource only prevents unitization from being a *requirement* to avoid litigation.

If pore spaces within aquifers are treated as a purely private property interest, any incidental use of storage by adjacent landowners could be considered a violation of their exclusive possession (redressed by actions in trespass, nuisance, conversion, or similar claims). At least in the oil and gas context, unitization agreements provide the voluntary consent of several landowners to the collective management of the reservoir – both its contents and its pore spaces.

7.3 A Brief Review of Law: Water and Property Rights in the Western United States

Recalling that water rights in the western United States originated from the practices of European mining and agricultural settlers, the first person in an area "staked a claim" to water they could use but lost those rights if they were unused over time. These users did not own the water but only had a usufructuary right. The birth of prior appropriation came from courts adopting the methods used by these European settlers. Courts, facing increasing water conflicts, could enforce the common law of England (which was used in the water-rich eastern United States) or incorporate a new common law legal regime specifically for the arid states. This common law system was eventually codified unto state statutes in many states. Groundwater rights emerged later, after technology allowed increased pumping. Groundwater rights, in a similar way, were inspired by the earlier precedents as models. Water storage rights were also drawn into the prior appropriation system for dams and diversions. Water-quality laws, however, did not rise until later, as environmental issues grew in prominence. A water right is simply a right to use a certain quantity of water from a certain source with no other aspects (quality, externalities, or ownership rights) of the water mentioned.

Groundwater rights are physically dependent on sediments and land, but that relationship is not represented in the water right. The rights associated with water are legally distinct from the property rights to the subsurface. Groundwater rights, while being a creature of water rights, have a direct influence on the pore spaces and fractures. Compaction of pore spaces leads to lost storage and subsidence, and real damages to real property owners. Is the compaction indirectly authorized by a water right? Or is the damage to the aquifer more akin to a tort, and subsidence induced by pumping damaging a real property right? If pore space rights are defined by the presence of water, they are a public property subject to state regulation. On the other hand, if pore spaces are creatures of real property rights, the use of pore spaces by groundwater users, ASR projects, carbon sequestration, and other public uses of aquifers become legally threatened. Water is a public good because of its critical value to life, while land and the subsurface are traditionally private goods. The state holds title to water resources as a component of its sovereignty, but it is unclear if that sovereignty extends to pore spaces holding that water. But these two resources are so physically interconnected as to render the distinction a legal fiction. For example, the Tribe's claim to the pore spaces below the surface of their reservation represents potential water storage, protection from subsidence, water quality improvement, and a meaningful expression of tribal sovereignty.

Property rights contain many "sticks" in the proverbial "bundle." Some of these rights include transfer, exclusive occupation, and use free from nuisance. Traditionally, these rights extend from the atmosphere above the parcel to the depths of the earth. Any minerals, sediments, and hydrocarbons are included with the property. Some rights to minerals or other materials may be severed from the property and held by someone else. Water beneath the earth, however, is owned by the public for its benefit. No water may be withdrawn without permission from a state actor.

Pore spaces containing water pose a conundrum. Private sediments (associated with a right of exclusion) are holding a public good (water in the filled portion of the aquifer). The public must have a right to privately held pore spaces (the gaps between privately held goods) for storage of public water. Otherwise, the public would be violating the exclusive ownership of the space below their land. No clear answer can be discerned from cases dealing with subsurface flooding, oil and gas trespass, or water contamination. Pore spaces are sometimes public and sometimes private, with no clear demarcation.

7.3.1 Pore Spaces: Defined by Their Contents?

Pore spaces cannot be easily differentiated from each other. Some contain water, some could contain water, some contain gas, and some contain oil. Legal regimes create rights associated with specific uses of resources. The law decides when something is a "pore space," "aquifer," or "storage space," "subterranean void," "hole," "pit," or "cave." The law also decides what rights are attached to each of those terms. While an expression of semantics, the choice will determine the ability of people to access and store water below their farms, cities, reservations, or public lands. Courts may have to decide what pore spaces count as aquifer, public pore spaces, or private storage rights and what (or if) those terms have rights associated with them. States have mixed approaches to pore spaces.

In cases of mineral laws and oil production, pore spaces are implicitly private property. In other cases involving water, pore spaces are treated as public property. When an aquifer is defined by the water contained within the pore spaces, the public rights associated with the water merge with the storage container. However, if the aquifer is all of the potential region that can store water, the public rights would also include any unfilled pore spaces and conflict with many other potential uses of the subsurface (like carbon sequestration, gas storage, water purification through infiltration basins, and subsurface wastewater cooling).

On the other hand, private pore spaces would complicate the clearly public waters being stored within them. Municipal ASR projects, wastewater treatment infiltration basins, and other public work projects could potentially occupy and contaminate private pore spaces on neighboring properties. This public use of private property would likely require compensation under the Takings Clause of the US Constitution. The cost of compensating these pore space owners could potentially end most public projects within aquifers.[1] Between these two extremes, a middle road should be found, protecting non-water private uses of pore spaces, while maintaining the public's interests in groundwater storage. A new kind of legal status, a public easement to groundwater storage, may serve as a route out of the quagmire.

7.3.2 Case Law for Pore Spaces as Private Property

The exact extent of any kind of property right is not easily read in a statute or clearly decided in a case. Property rights are a creature of common law and practice and are subject to modern revisions. Rarely does a court identify the extent of property rights, but implicitly recognize the right when a landowner brings claims for trespass, conversion, takings, or nuisance. Courts only face the question of whether a property right exists when a party brings a claim of injury from the actions of another under tort law. Therefore, the outline of the extent of property rights in the subsurface can come to light through a review of these kinds of claims.

The classic case defining property rights is the *Causby* case.[2] The case took place in 1946 when the negative impacts of aircraft were just beginning to

---

[1] See DWA Summary Judgment Phase 2 at 8 (stating "Under the Tribe's ownership theory, MWD and other public water agencies that store water in groundwater basins as part of their conjunctive use programs would be required to pay compensation to those who own the lands where the water is stored.")

[2] *United States v. Causby*, 328 U.S. 256, 260 (1946) ("*Causby*").

be felt. One of these impacts is the noise associated with engines near airports. For a rural chicken farmer, these noises can lead to real issues. The *Causby* case began when aircraft near a naval air base were so noisy that Thomas Causby's chickens were dying in their pens.[3] The farmer claimed that the aircraft were flying in his property, since he owned the atmosphere above the land.[4] The bombers and fighters were flying within twenty feet of the tops of trees on his property, generating loud engine noises and lights near his home and his chickens. The US Supreme Court took the case to determine if the aircraft were actually occupying and taking his property for public use.[5] The ancient principle of ad coelum was the starting point for the extent of private property ownership.[6] The phrase translates to "whoever owns [the] soil, [it] is theirs all the way [up] to Heaven and [down] to Hell" (Hepburn, 2014, pp. 10310, 10313). Applied to this case, it would have meant that the aircraft, or any aircraft flying in any airspace, were trespassing at the least or stealing property at the worst. However, the Supreme Court would not allow the advancement of aircraft technology to be stopped by strict application of an ancient doctrine. The Supreme Court made an exception for the extremes of height, since there would be no practical way for a landowner to use all of the atmosphere above their property.[7] At some level, the atmosphere becomes a public highway, available to all.[8] Since the aircraft were flying so low, the Supreme Court remanded the case to see if damages would be proper.[9]

Like the region of the atmosphere above a property, the soil of a property is also a component of the ownership rights. The ad coelum doctrine has been incorporated into arguments for aquifer rights, but with mixed results. The presence of water often sways the court in favor of finding a public right or convinces the court that water law should determine the outcome. When water is not involved, the courts are more willing to find that a private property right exists. The more modern cases that find a private right to pore spaces typically arrive from mineral, oil, and gas cases.

When material is removed during oil, gas, or other mineral extraction, courts typically find that the pore spaces are the property of the overlying landowner. For example, when mineral rights are severed from the overlying property owner, the rights to the minerals may be owned by another person. However, the spaces

---

[3] Ibid. at 258.
[4] Ibid. at 259.
[5] Ibid. at 259–260.
[6] Ibid. at 260.
[7] Ibid. at 265.
[8] Ibid.
[9] Ibid. at 268.

holding that material typically remain with the landowner. In multiple case, courts have found the mineral rights cease after the fluids have been extracted.[10]

While pore spaces are quite small in size, larger voids have been clearly identified with the surface owner, not the owner of the stored material, like oil or water. For example, large "jugs" were created in salt deposits to store oil under marshes in Louisiana.[11] Water was used to erode the salt deposits to create containers that could store over a million barrels of hydrocarbons.[12] The question for the court was who owned the voids, the landowner or the mineral estate owner.[13] These subterranean voids were found to be the property of the landowners, not the mineral estate owners.[14] The court states, "It would surely be absurd for the mineral owner to then assert that he has acquired certain rights to the hole in the ground made by his mining operations." But does this legal process apply to naturally occurring holes in the ground?

Natural formations capable of storing gases have also been found to be the overlying landowner's property.[15] The Bush Dome is a region where natural gas was collected as early as the 1920s. The formation has the potential to store fifty-two billion cubic feet of gases.[16] The conflict arose when the owner of gas rights wished to store helium within the formation rather than only extract natural gas.[17] The rights to extract gas did not include the right to store foreign materials in the formation, even when helium was naturally occurring in the formation.[18] As this case shows, natural storage spaces are also the property of overlying landowners. But does this property right apply in aquifers, or is it only in mineral rights issues?

At least in Texas, the extraction of water cannot injure the storage rights of neighboring landowners.[19] In this case, water extraction caused substantial subsidence of a neighborhood.[20] The subsidence caused a lake to flood into their land and their homes.[21] The court decided that the neighborhood did have a right to the supportive pressures within the pore spaces below their property. The court states, "It appears that the ownership and rights of all landowners will be better protected against subsidence if each has the duty to produce water from his land in a manner that will not negligently damage or destroy the lands of others."[22]

---

[10] See *DOT* v. *Goike*, 220 Mich. App. 614, 616 (1996); *United States* v. *43.42 Acres of Land*, 520 F. Supp. 1042, 1042 (W.D. La. 1981).
[11] *United States* v. *43.42 Acres of Land*, 520 F. Supp. 1042 at 1043.
[12] Ibid.
[13] Ibid. at 1044.
[14] Ibid. at 1046.
[15] *Emeny* v. *United States*, 412 F.2d 1319, 1323 (1969).
[16] Ibid. at 1321.
[17] Ibid. at 1325.
[18] Ibid.
[19] *Friendswood Dev. Co.* v. *Smith-Sw. Indus.* 576 S.W.2d 21, 21 (Tex. 1978).
[20] Ibid. 21–22.
[21] Ibid.
[22] Ibid. at 30.

It appears from this case that groundwater pumpers are also responsible for preventing pore space compactions in aquifers. Only a private right to pore spaces within an aquifer could provide support for the claim that the court recognized.

Three states have explicitly stated that a private pore space right is recognized by law.[23] The Montana statute does not mention pore spaces, but states that "storage reservoirs" are property of the overlying owner unless specifically severed.[24] A Wyoming statute gives landowners the exclusive ownership right to storage of carbon or "other substances" within the "pore spaces" under the surface of the earth.[25] North Dakota also grants ownership rights to "pore space in all strata" to the surface landowner.[26] These statutes are not specific to what materials, what depth, or what history of uses have occurred.

Issues with laws like the ones described earlier have already resulted in controversy. In *Mosser* v. *Denbury*, 898 ND 169 (2017), the North Dakota Supreme Court decided that landowners must be compensated when adjacent mineral developers deposited waste saltwater related to oil and gas operations into pore spaces beneath the land of others. Citing North Dakota Century Code 47-31-03, the Supreme Court affirmed that the meaning of "land" under North Dakota Century Code 38-11.1-01 includes pore spaces and requires compensation to a surface landowner when used by a mineral developer for wastewater disposal. In response, the North Dakota legislature passed Senate Bill 2344 (2019), which created North Dakota Century Code 47-31-09 and altered section 38-08-25, section 38-11.1-01, and section 38-11.1-03. These legislative actions relieved mineral developers from the requirement to compensate landowners when wastewater intrudes into pore spaces beneath their property. These statutory changes are currently being challenged by a group called the Northwest Landowners Association, which is seeking to revert the changes created by Senate Bill 2344 (2019). As of this writing, the outcome has yet to be determined – whether these statutory changes constitute a "taking" of private property, requiring government compensation. These kinds of controversies, however, appear to be limited to oil and gas operations and have yet to be addressed to aquifers.

7.3.3 Pore Spaces as Public Ownership

Courts have recognized public rights to pore spaces just as implicitly as some courts have decided pore spaces are protected private rights. These cases tend to

---

[23] See M.C.A. 82-11-180 (Montana); W.S.A 34-1-152 (Wyoming); N.D.C.C 47-31-04 (North Dakota).
[24] M.C.A. 82-11-180.
[25] W.S.A 34-1-152(a)-(d).
[26] N.D.C.C 47-31-04.

originate in water law issues, dealing with migrating contamination or subsurface flooding. The public character of the water also appears to flow into the pore spaces that hold the groundwater. For the Agua Caliente litigation, the groundwater claims in the same lawsuit may serve to sway the court toward finding a public right to the reservation's pore spaces. As seen in the cases discussed later, water complicates an otherwise easy finding by many courts.

The first example shows that courts favor public rights to store water under the land of others.[27] A statute allowed water right holders to claim water that incidentally leaked from ditches into the surrounding aquifer.[28] The landowners claim that the statute effectively took their rights to store water below their property and gave that right to other water users.[29] But the court did not rule under property law principles. Instead, the court reasoned that the statute did not interfere with the landowner's current uses of groundwater, nor with the landowner's right to exclude (a property law right).[30] The court, when faced with a claim under property law, chose to rule under water law within an aquifer pore space.

Another court also ruled that an ASR project does not invade the property of others, because it does not require the use of a "man-made facility."[31] Under Colorado law, taking of property requires compensation when man-made water storage facilities are placed on private property.[32] While the ASR system was a man-made facility, no compensation was required for filling pore spaces of nearby properties because it was a natural feature of the landscape.[33] The court, under this logic, would also not compensate land submerged after the construction of a dam, but this would clearly not be the predicted outcome. This unsteady logical leap shows the extent to which some courts will go to try to avoid the question of finding a property right in an aquifer.

With the courts entrenched in opposing sides of the argument, no clear answer can be given for property rights in aquifers and pore spaces. The cases, however, divide cleanly on whether the pore spaces are filled with water or some other substance. While state statutes and mineral law cases point to private ownership of the pore spaces, the public aspects of water resources sometimes imply public ownership of the aquifer as well. Courts have not often clearly identified the legal status of public use, but a few have been able to bridge the gap.

---

[27] *In re Application U-2*, 226 Neb. 594, 598 (1987).
[28] Ibid. at 596.
[29] Ibid.
[30] Ibid.
[31] *Bd. of Co. Comm'rs v. Park Co. Sportsmen's Ranch, LLP*, 45 P.3d 693, 693 (2002).
[32] Ibid. at 708.
[33] Ibid. at 713–714.

7.3.4 A Mixture of Both?

For public and private rights to exist in the same space, common law provides multiple approaches for providing landowners with adequate rights while allowing others access. Some of these are private law, like covenants and easements. Other examples come from public sources, like the public trust, reserved water doctrines, and navigation servitudes. These doctrines allow two (or more) parties to make use of the same space with cleanly defined rights and obligations between those parties. Because groundwater aquifers are unlike any other resource, novel approaches are required. These mixed solutions would provide aquifers with clearer lines and bring order to the courts' treatment of pore spaces. At least one court has applied this approach to aquifers under state law.

The potential conflicts in the future for aquifers will be about filling, not draining, the pore spaces. One such case is the first to suggest a compromise could be made between the two extremes of ownership. In *Niles Sand & Gravel Co.* v. *Alameda County Water Dist.*, a California court faced the question of whether an ASR project had a right to fill pore spaces, or whether a gravel pit operation had a right to be free from flooding caused by the project.[34] An ASR project was injecting water into the aquifer to prevent saltwater intrusion from the sea.[35] The locally increased water table filled the gravel pit. It required extensive pumping to remove the water.[36] The pumped water was considered "waste" and was dumped into the San Francisco Bay.[37] The gravel pit owner claimed that the ASR project was a public taking of their pore space property rights, while the ASR project claimed that the gravel pit harmed their saltwater barrier efforts.[38] The court eventually found that California has recognized a "public servitude" to a certain water table depth since the creation of the state and adoption of the correlative rights system.[39] The gravel pit did not have a property right that was infringed upon, but it was actually interfering with an easement-like right of use by the public. The gravel pit was denied any compensation for the public's use of pore spaces within their property.[40]

A public servitude to pore spaces does not surrender all rights to a property, only the use of those pore spaces that contain water under the correlative doctrine. In addition, the servitude would only apply up to the level of the water table. The servitude approach also would not apply to deeper pore spaces, like

---

[34] 37 Cal. App. 3d 924, 926 (1974).
[35] Ibid. at 930.
[36] Ibid. at 926.
[37] Ibid.
[38] Ibid. at 932–933.
[39] Ibid. at 935.
[40] Ibid.

Table 7.1. *Examples of cases with summaries that address pore space ownership*

Case citation	Implicitly public, private, or mixture?	Summary of approach to pore space ownership
Niles Sand & Gravel Co. v. *Alameda County Water Dist.*, 37 Cal. App. 3d 924, 926 (1974)	Public	An ASR project flooded a mining pit, requiring additional pumping to dewater the pit. The court considered the pumping "waste" of water and ruled against the mining pit owners.
Los Angeles v. *San Fernando*, 14 Cal. 3d 199, 256 (1975)	Public	A municipal ASR project injected water into an aquifer. The case omits discussion of the ownership of the aquifer and focuses on the ownership of recharged groundwater migrating to adjacent jurisdictions.
Friendswood Dev. Co. v. *Smith-Sw. Indus.*, 576 S.W.2d 21, 21 (Tex. 1978)	Mixture	Subsidence caused surface flooding, indirectly related to pore space compaction. The court applies nuisance theory to the use of water rather than recognizing a property interest in the aquifer.
Nunez v. *Wainoco Oil & Gas Co.*, 488 So. 2d 955 (La. 1986)	Private	Subsurface drilling into adjacent property at a depth of 2 miles supported claims for trespass.
In re Application U-2, 226 Neb. 594, 598 (1987)	Public	The court stated the plaintiff landowner had no standing to challenge seepage from a canal nearby that filled pore spaces of adjacent property, since no injury occurred (failing to recognize a right to exclude).
Bd. of Co. Comm'rs v. *Park Co. Sportsmen's Ranch, LLP*, 45 P.3d 693, 693 (2002)	Public	An ASR project planned to recharge an aquifer, filling pore spaces of adjacent property owners. Statutes authorizing the use of aquifer for storage supplanted the landowners' common law theory of ownership of pore spaces. The analysis applied water law principles rather than property law principles.
Mosser v. *Denbury*, 898 ND 169 (2017)	Private	Members of a unitization agreement were unable to raise claims of nuisance and trespass for imported saltwater into pore spaces below their property, since

Table 7.1. (*cont.*)

Case citation	Implicitly public, private, or mixture?	Summary of approach to pore space ownership
		it appeared to be authorized by the unitization agreement. However, the landowners must be compensated for the lost economic value associated with the lost use of the pore spaces under a state statute.

those used in the oil and gas industry or with carbon sequestration. The balanced approach of a public pore space right of use allows for both public and private uses of pore spaces.

7.3.5 Pore Spaces: A Void in the Law

Someday, court may settle on a more universal approach to pore spaces. We have reviewed many cases, summarized in Table 7.1 that, for one reason or another, appear to confuse water law with property questions, merge principles from separate areas of the law, and generally avoid answering the question of the scope of pore space ownership directly. We argue that these kinds of controversies will become even more frequent and require a more standard and generally accepted approach. A review of the cases in Table 7.1 should provide a fruitful (if not confusing) review of the status of the ownership of aquifer pore spaces.

While we have provided a general review of some courts' views on this matter (directly or implicitly), we encourage the reader to keep a keen eye on this area of the law as it develops in the future. To truly see how this question could be resolved, we suggest digging into our summary of the arguments, positions, and analysis of the Agua Caliente litigation in the next chapter. Like many of the cases discussed earlier, no clear answer may come from the litigation. However, it may be one of the first glimpses into a court that directly engages with this important issue.

8

Pore Spaces

The Agua Caliente Litigation

The Agua Caliente Band of the Cahuilla Indians, a historic Palm Springs–based tribe in California, pursued the establishment of a right to the pore spaces, not the water alone, in the aquifer below their reservation (see also Chapter 2).

The litigation has multiple legal approaches and stages. First, the Tribe wished to establish their right to use and protect the groundwater below their reservation. Second, the Tribe wished to assert their rights to the pore spaces of the aquifer, which would expand their rights to store water in the aquifer and offer protection from subsidence due to off-reservation groundwater pumping. Unfortunately, as discussed later, the courts may not provide a clear answer due to a lack of "standing" for the Tribe to assert its claim to pore spaces. But, as no other case appears to directly address this question, a detailed examination of the case could provide some clarity on the shape these debates will take in the future.

The following portions of this chapter review the stages and arguments in the Agua Caliente litigation. First, we will briefly describe aquifers and the distinction between groundwater and pore spaces in legal and hydrogeological senses. Next, we will describe the current Tribe's arguments regarding reserved water rights and the first phase of the litigation. We will then go over the second phase of the litigation, which asserts rights to pore spaces and a specific water quality, and identifies the legal standard for the quantification of tribal groundwater rights. Finally, the implications on various outcomes of the ligation are explored, showing the scope the litigation could have on aquifer management across the country. Because the litigation necessarily has to make choices balancing public and private rights to aquifers, it may be one of the first court cases to attempt to directly resolve these questions. The final portions of this chapter will focus primarily on the pore space aspects of the litigation, developing the issues and inconsistencies of groundwater and property law.

8.1 Agua Caliente Litigation and Its Potential to Flood Out the Indian Law

Like the way Native American rights to surface water eventually led to the establishment of federal rights to water, the outcome of the decision on pore spaces rights (if decided in its merits) could have implications on non-Native American rights to pore spaces in aquifers. If found to be the property of surface owners, pore space ownership complicates the legal foundations for multiple emerging technologies, like ASR and carbon sequestration. The Agua Caliente litigation has the potential to expose assumptions of public rights to pore spaces used by these kinds of projects. The extent of public rights to a groundwater aquifer's pore spaces will need to be balanced against the private interest in protecting the use of stored water from an ASR project.

The court will be forced to decide issues that are not specific to Native American reservations but to non-tribal aquifers in many states. Just as the *Winters* doctrine eventually led to the development of the federal reserved water right doctrine, the decisions about reserved rights to groundwater, water quality, and pore spaces could spread to any federally held project or public lands. The influence could reach further, and expose issues within state law that were previously ignored. Water quality has not been associated with groundwater rights under state laws. Water rights do not specifically state what kinds of groundwater count as "water." Does it include brackish groundwater found at extreme depths or produced water as by-product of hydrocarbon production? Courts may be forced to decide what kinds of groundwater are intended by a federally reserved water right.

8.1.1 History, Hydrology, Rights, and the Coachella Valley

Before the arguments regarding groundwater and pore spaces claimed by the Tribe can be discussed, the history and hydrology of the Coachella Valley can provide context to the litigation. While distinctions between sources of law and their influences will be reviewed later in this chapter, we will review the movement of water caused by physical forces and human ingenuity. The Tribe's historical water uses and the recent increases in groundwater withdrawals influence the earth on which the reservation rests and the future lives of the Tribe's members. The physical and human context defines the issues and forces behind the litigation.

8.1.2 The First Water Users: The Tribe

The Tribe has used the groundwater in the Coachella Valley effectively forever. The Coachella Valley lies in an arid portion of Southern California. Surface water is available, but only in limited quantities.[1] Multiple rivers, including the Whitewater River, flow through the Coachella Valley and terminate in the Salton Sea.[2] The valley forms a "sink" where water does not flow to the ocean, meaning any precipitation in the basin either evaporates or percolates into aquifers.[3] Despite limited access to water, the Tribe has historically had extensive agricultural activities in the valley. The Tribe has grown grains, vegetables, and fruits in the valley using surface water ditches, and the Tribe may have acquired water from thirty-foot deep hand-dug wells within settlements.[4] Accounts from early European settlers describe the Tribe's agricultural use of a mile-long irrigation ditch as early as 1830s.[5] Water use by the Tribe in the valley has likely taken place for many centuries.

Surface water has been reserved by the Tribe since 1938, when the Superior Court of the State of California, Riverside County, adjudicated the Whitewater River.[6] The Tribe was granted 6.0 cubic feet per second (cfs)[7] with a priority date of January 1, 1893, and available throughout the entire year. Additionally, the adjudication granted the Tribe rights to Tahquitz Creek for 4.8 cfs, available the entire year, with a priority date of April 26, 1884.[8]

With little water naturally available in the basin, canals provide imported water from the Colorado River, which is stored in the aquifer using ASR technology. With limited access to surface water, groundwater and the pore spaces storing it are a critical resource to the Tribe.

8.1.3 The New Users: Water Districts and the State Water Project

The Tribe's claims for groundwater originate from a need to protect their future and the development of groundwater for the reservation. The nearby Desert Water Agency (DWA) and Coachella Valley Water District (CVWD) are the major water development and management organizations in the region. These organizations coordinate under the California State Water Project (SWP) to exchange water with the Metropolitan Water District (MWD).[9] The two agencies operate a number of groundwater replenishment systems using imported Colorado River water.

---

[1] See Complaint at 5–7.
[2] Ibid. at 8–9.
[3] https://pubs.usgs.gov/fs/2007/3097/pdf/fs20073097.pdf
[4] Tribe Complaint at 6. However, the Tribe was unable to provide evidence of these wells in discovery; see CVWD Summary Judgment at 13.
[5] Tribe Complaint at 6.
[6] Ibid. at 10.
[7] There are 35.3 cfs in one cubic meter per second.
[8] Tribe Complaint at 10.
[9] Desert Water Agency Fact Sheet at 1 (see https://dwa.org/board-meeting-agenda/misc/107-imported-water-fact-sheet/file).

The MWD delivers water from the Colorado River Aqueduct into the Whitewater River. The delivered water is exchanged for the CVWD and DWA's water available under the SWP.[10] California SWP water is gathered from the Sacramento River system, but the canal works do not reach the Coachella Valley.[11] The Colorado system does reach the valley, so the agencies trade their volumes available from the SWP to MWD, which has access to the Colorado River.[12] The delivered Colorado River water flows in the Whitewater River to spreading basin groundwater replenishment systems operated by CVWD and DWA. The Colorado River water has a significantly higher dissolved solid content than groundwater naturally occurring in the Coachella Valley.[13] For example, the salinity of naturally occurring local groundwater is approximately 130–2,000 parts per million (ppm). Imported Colorado River water has a salinity of approximately 550–750 ppm.[14] The safe level for drinking water is 1,000 ppm for salinity.[15]

8.2 The Progress of the Litigation

For the Tribe to secure access to groundwater, protect water quality, and protect its ability to store water below their reservation, a court must find that the federal government reserved a portion of groundwater for establishing the Tribe's homeland. The claim for pore spaces, however, is not established under the same legal doctrines as claims for groundwater. While the two resources are related, as discussed earlier, the legal principles are distinct. To make their case, the Tribe filed a complaint against the CVWD and DWA on May 14, 2013.[16]

8.2.1 The Opening Salvo: The Complaints, Answers, and Trifurcation

In the complaint initiating the legislation, the Tribe seeks a "declaration of the Tribe's right to use pore space in the aquifer underlying the Coachella Valley to store the Tribe's federally reserved water in an amount sufficient to meet all of the Tribe's present and future reasonable needs."[17] To support their claims, the complaint lists "irreversible subsidence, decreases in groundwater quality, declining water levels and increased water extraction costs."[18] The claims for relief include declarations by the court of the Tribe's federally reserved

[10] MWD Amicus Brief at 2.
[11] CVWD Summary Judgment Phase 2 at 6.
[12] Ibid.
[13] Tribe Complaint.
[14] CVWD Salinity Fact Sheet; see www.cvwd.org/DocumentCenter/View/63
[15] Ibid.
[16] See Tribe Complaint at 1.
[17] See Complaint Filed May 14, 2013, p. 4.
[18] Ibid. at 13.

groundwater rights, a right to a certain level of water quality to fulfill the purpose of the groundwater right, and an "ownership interest" in the pore spaces and their "groundwater storage capacity" in the aquifer. The Tribe claims that the replenishment of the aquifer with Colorado River water introduces increased amounts of "TDS [total dissolved solids] and salinity," damaging the existing groundwater.[19] The United States intervened in the litigation on June 19, 2013.[20] The complaint for the United States includes arguments for a federally reserved water right and advocates for recognition of an implicit water quality standard and the use of the "homeland" quantification method. The complaint for the United States omits the claims for ownership of pore spaces alleged in the Tribe's complaint. The Tribe, however, has maintained its claims to ownership of the pore spaces throughout the litigation.

The answers submitted by the CVWD and the DWA denied most substantive claims outlined in the Tribe's complaint. The CVWD and DWA answer claims that the pore spaces are not a private ownership interest but a "public resource" available to all.[21] The agencies do not cite any authority for their position, but their argument is at the heart of the debate over aquifer–groundwater right distinctions.

Because of the complexity of the arguments raised, the parties stipulated to trifurcate the litigation. In the first phase of the litigation, the court decided if the Native American federally reserved right to water extends to groundwater and if the Tribe has aboriginal rights to groundwater under the reservation. In the second phase, the court will decide if the right includes a right to a certain water quality, if there are ownership rights in the pore spaces of the aquifer, the standard of deciding the quantity of water reserved, and the defenses posed by the DWA and CVWD.[22] In the final phase, the court will decide the fact-intensive quantification of water quantity and quality, the volume of pore spaces owned by the Tribe, and injunctive remedies needed.[23]

8.2.2 The Phase 1: Summary Judgments and Trial Court Opinion

The first phase of the litigation merely addresses the foundational question on whether the Tribe has a right to groundwater. Under the *Winters* line of cases, Native American tribes have an implicitly created right to water for the support of

---

[19] Ibid. at 14.

[20] Paperless motion on June 12, 2013. Docket 5:13-cv-00883.

[21] Citing *Central and West Basin Water Replenishment Dist.* v. *Southern California Water Co.*, 109 Cal. App. 4th 891 (2003).

[22] *Agua Caliente Band of Cahuilla Indians* v. *Coachella Valley Water Dist.*, No. EDCV 13-883-JGB, 2015 U.S. Dist. LEXIS 49998, at *6 (C.D. Cal. March 20, 2015); Stipulation to Bifurcate Case December 2, 2013.

[23] Ibid.

the purposes of the reservation, such as for agriculture.[24] The priority date for a reserved water right is typically established on the date of the creation of the reservation.[25] *Winters* rights are superior to state-law-created water rights and are not subject to restrictions imposed by state law.[26] A later case established that the quantification method would be determined by the "practicably irrigable acres" of land located on the reservation.[27] However, other courts have deviated from this quantification method, using a "homeland" standard. In addition to the *Winters* doctrine, Native American tribes may attempt to establish aboriginal title to water.[28] Using this method, a treaty between the Tribe and the United States is analyzed to determine if the treaty diminished the "right of occupancy" held by the Tribe.[29] Water rights established with aboriginal title use a time immemorial priority date, becoming the earliest priority date in the basin.[30]

All four parties to the litigation filed summary judgments on whether the Tribe could claim rights to groundwater in the same way as surface water. The Tribe's summary judgment claimed that the reservation had a federally reserved right and aboriginal title to groundwater.[31] The Tribe's argument for groundwater relies on the rationale behind the *Winters* doctrine. The Tribe claims the *Winters* doctrine applies with "equal force" to surface and groundwater.[32] The Tribe relies on the *Cappaert* decision, which addressed the doctrine's application to a small pool filled with groundwater that contained an endangered fish.[33] The appellate court treated the pool as groundwater, and supported the equal application of the *Winters* doctrine to either source. However, the Supreme Court did not define the pool as groundwater but stated "[h]ere, however, the water in the pool is surface water."[34]

For the Tribe's aboriginal title claims in their summary judgment, the Tribe claimed that it had established rights to groundwater through long use and occupancy of the reservation lands.[35] While the statute governing land claims after the Mexican–American War required all claims to be registered before 1853,[36] the Tribe alleges that its aboriginal rights were not subject to the Mexican or United States statutes.[37] The Tribe claims, instead, that the rights were established since

---

[24] *Winters* v. *United States*, 207 U.S. 564, 576 (1908).
[25] See Ibid. at 576–577.
[26] See Ibid.
[27] *Arizona* v. *California*, 373 U.S. 546, 598–600 (1963).
[28] *United States* v. *Adair*, 723 F.2d 1394, 1414 (9th Cir. 1983).
[29] See Ibid.
[30] Ibid.
[31] See *Winters* v. *United States*, 207 U.S. 564 (1908).
[32] Tribe Summary Judgment at 14.
[33] See ibid. at 15.
[34] *Cappaert* v. *United States*, 426 U.S. 128, 142 (1976).
[35] See *Sac and Fox Tribe of Indians of Okla.* v. *United States*, 383 F.2d 991, 998 (Ct. Cl. 1967).
[36] *An Act to Ascertain and Settle the Private Land Claims in the State of California.* 9 Stat. 631 (March 3, 1851) ("Registration Statute").
[37] Tribe Summary Judgment at 21.

time immemorial by use and occupation by the Tribe.[38] Further, the statute states that it does not authorize settlement on lands occupied by Native American tribes.[39] Finally, the Tribe asserts that any rights that were lost due to the lack of registration were reestablished.[40] The United State's summary judgment echoes the *Winter's* doctrine arguments, but does not repeat the aboriginal title arguments.

The Tribe's claims were attacked by both agencies in their summary judgments. The DWA's summary judgments argued that the Tribe has only rights established under state law for groundwater.[41] The DWA argued that the reserved rights are limited to the amount necessary to prevent the entire defeat of the purposes for the reservation.[42] Since California already provides groundwater rights for the reservation, the DWA argues that a federally reserved right is unnecessary.[43] Since there is no priority system for groundwater in California, the DWA argued that there is no conflict between federal interests and state laws.[44] The Tribe has rights established under state law that it chooses not to exercise and is not prevented from accessing groundwater within the reservation. In addition, the DWA argued that applying a federally reserved water right, possibly defined by prior appropriation, would create the kind of confusing water regime identified by the *California* and *Walton* courts.[45] The ability of DWA to effectively manage water of the Coachella Valley aquifers would be hindered by state and federal water right regimes operating within the same aquifer. The DWA alleged that all necessary water had been provided in the adjudication of the Whitewater River in 1938.[46] The DWA also attacked the Tribe's aboriginal claims, providing precedent that the Registration Statute extinguished these claims.[47] Because the Registration Statue extinguished any unregistered claims to land, any claims to water would also have expired in 1853. Similarly, the summary judgment for the CVWD dismissed the aboriginal title arguments, citing the Registration Statute. In addition, CVWD argued that the *Cappaert* decision was based on surface water reservations, not groundwater, unlike the Tribe's characterization of the case.[48]

Ultimately, the trial court decided to find a federally reserved right for the Tribe's access to groundwater, but did not accept the Tribe's arguments concerning

---

[38] Ibid.
[39] Ibid. at 22.
[40] Ibid. at 23.
[41] See DWA Summary Judgment at 6.
[42] Ibid. at 7.
[43] Ibid. at 15.
[44] Ibid. at 8.
[45] Citing *California* v. *United States*, 438 U.S. 645, 653–654 (1978) and *Colville Confederated Tribes* v. *Walton*, 647 F.2d 42, 53 (9th Cir. 1981), which states "Congress almost invariably defers to state water law when it expressly considers water rights."
[46] See DWA Summary Judgment at 24.
[47] Citing *Barker* v. *Harvey*, 181 U.S. 481 (1901).
[48] CVWD Summary Judgment at 20.

aboriginal rights to groundwater.[49] Because the Tribe did not make a claim under the Registration Statute, made no claim in the intervening years before creation of the reservation in 1876, the Tribe effectively recognized the United State's title to the reservation lands.[50] The creation of the Tribe's "reservation effectively re-extinguished that right."[51] The court categorized the main arguments of the agencies as a federalism argument for state water law, and another for asserting that California water law provides adequate water to fulfill the purposes of the reservation.[52] The court decided that these issues relate to quantification questions, not the mere existence of a right addressed in phase 1 of the trial.[53] The court granted the summary judgments for the Tribe and the United States in favor of finding a federally reserved right to groundwater, but granted summary judgments to the agencies, failing to recognize an aboriginal right to groundwater.[54]

The agencies filed interlocutory appeals of the trial court's phase 1 decision to the United States Court of Appeals for the Ninth Circuit. In response to a motion to stay the litigation during the appeal, the Tribe opposed the stay and wished to continue on the determination of the legal standard of quantification of the groundwater right and if the Tribe has an ownership interest in the pore spaces in the aquifer.[55] The Tribe argued that issues related to pore spaces are distinct from reserved groundwater rights, and could be decided in the intervening time while the groundwater issues were reviewed.[56] Because the potential groundwater rights would have no bearing on the pore space property interests, these could be litigated. The court, however, disagreed with the arguments and stated, "Plaintiffs have not convinced the Court that the issue of ownership of 'pore space' is separate and distinct from the Tribe's water rights."[57] The court approved the stay on all issues, since the court felt that the entire case rested on the question of groundwater rights.[58]

[49] *Agua Caliente Band of Cahuilla Indians* v. *Coachella Valley Water Dist.*, No. EDCV 13-883-JGB, 2015 U.S. Dist. LEXIS 49998, at 36 (C.D. Cal. March 20, 2015).

[50] Ibid. at 33.

[51] Ibid.

[52] Ibid. at 22.

[53] Ibid.

[54] Ibid.

[55] See Joint Report filed on May 14, 2013 at 6.

[56] See Report on May 11, 2015, at 6, stating "[o]n the contrary, Agua Caliente owns the pore space beneath its Reservation regardless of whether it also has a federal reserved groundwater right. See, e.g., *United States* v. *Shoshone Tribe of Indians of Wind River Reservation in Wyo.*, 304 U.S. 111, 116 (1938) (holding, in a case not involving water rights, that tribal ownership of reservation lands includes ownership of all 'constituent elements of the land itself,' including subsurface rights); *United States* v. *43.42 Acres of Land*, 520 F. Supp. 1042 (W.D. La. 1981) (holding that the owners of the surface own and are entitled to compensation for 'the underground storage value of the land'); accord *Starrh & Starrh Cotton Growers* v. *Aera Energy, LLC*, 63 Cal. Rptr. 3d 165, (Cal. Ct. App. 4th 2007) (characterizing groundwater pore space – not groundwater itself – as a mineral estate owned by the landowner under California law)."

[57] See Order on September 8, 2015, at 4.

[58] Ibid. at 5.

Even when the Tribe argued cases entirely distinct from reserved groundwater rights, the court took the position that no rights to pore spaces exist if no rights to groundwater are found. While this position ignores the property-based arguments in the Tribe's motion, it shows that the court may later decide in phase 2 that pore space storage is a water law, not property law, issue. While this may seem at first glance to support public rights to aquifers, it would also considerably undermine the Tribe's ability to utilize and protect the groundwater rights it may be granted. The issue of pore spaces, unfortunately, waited until after the completion of the interlocutory appeal.

8.2.3 *Appellate Opinion from the Ninth Circuit*

The court of appeals affirmed the decision of the trial court on all issues.[59] The court of appeals summarized the arguments of the agencies by looking at their focus on their reliance on necessity of water for the purposes of the reservation.[60] The agencies argued that in determining whether there was a reserved right, the court must look to the purpose of the reservation and determine if water was necessary for that purpose.[61] In addition, the agencies alleged the court must also determine if water presently available under state laws is insufficient before deciding if water was reserved.[62] The court rejected both of these contentions, simplifying the question to if, as a matter of law, a tribe may have a reserved right to groundwater.[63] The questions of state law rights and purpose are quantification questions not related to the threshold issue.[64] In addition, any current uses of water do not matter to the determination of a reservation to fulfill the purposes of the reservation.[65] The court, in affirming the trial court's opinion, also suggested that the trial court should find some amount of water reserved in the aquifer.[66] After the appellate court issued its opinion, the work began on phase 2 of the litigation. This case may be the first time a federal appellate court clearly and explicitly recognized a Native American reserved right to groundwater. On November 27, 2017, the United States Supreme Court denied review of the decision of the court of appeals, effectively resolving any issues presented with phase 1 in favor of the Tribe's reserved right to groundwater.

---

[59] *Agua Caliente Band of Cahuilla Indians* v. *Coachella Valley Water Dist.*, 849 F.3d 1262, 1273 (9th Cir. 2017).
[60] Ibid. at 1269.
[61] Ibid.
[62] Ibid.
[63] Ibid. at 1270.
[64] Ibid.
[65] Ibid. at 1272.
[66] Ibid. at 1273, stating, "Thus, to guide the district court in its later analysis, we hold that the creation of the Agua Caliente Reservation carried with it an implied right to use water from the Coachella Valley aquifer."

8.2.4 *Phase 2: Summary Judgments*

With the decision to trifurcate the litigation, the second phase of the trial would deal with potential pore space ownership, a water quality implicit to any reserved water right, the standard to be applied for quantification, and any other defenses raised by the agencies.[67] All parties filed motions for summary judgment on these threshold issues on October 20, 2017. The various arguments have been summarized and compared in the following sections.

8.2.4.1 *Pore Spaces*

The arguments over the ownership of pore spaces in the aquifer take two broad forms: a public resource available to all or a private property right. The agencies took the position that pore spaces were a public resource. The CVWD argued in their summary judgment that the issue was not ripe or lacked standing for litigation.[68] Because the Tribe was not currently withdrawing or planning to withdraw groundwater, there was no damage to the potential storage rights of the Tribe.[69] The approach for this argument relies on pore spaces being directly related to any groundwater rights established by the Tribe. The CVWD also argued that the question of ownership of the pore spaces lacked standing.[70] CVWD's reasons for attacking the Tribe's standing on this issue were similar to the arguments regarding storage rights. The Tribe could only show a minor effort at groundwater recharge and no plans for further use of the pore spaces.[71] Therefore, there was no controversy for the Tribe to bring the claim to court.

The CVWD further argued that aquifer storage spaces are considered a public property. The argument relies on cases that only address groundwater, not mineral rights.[72] These cases draw from water rights doctrine, showing that the public ownership of storage space can be inferred from the presence of groundwater (a public resource).[73] The right to withdraw water came with a right to store water in the aquifer.[74] In one case, the pro rata division of pore spaces to overlying landowners would fail to support the public's beneficial use of the pore spaces.[75]

---

[67] See Order on September 3, 2015, at 2.

[68] CVWD Summary Judgment Phase 2 at 13.

[69] Ibid.

[70] Ibid.

[71] Ibid. at 13–14.

[72] Citing *Bd. of Cnty. Comm'rs of Cnty. of Park* v. *Park Cty. Sportsmen's Ranch, LLP*, 45 P.3d 693, 696 (Colo. 2002); *Cent. & W. Basin Replenishment Dist.*, v. *S. Cal. Water Co.*, 109 Cal. App. 4th 891, 905 (2003); *City of Los Angeles* v. *City of Glendale*, 142 P.2d 289 (Cal. 1943); *City of Los Angeles* v. *City of San Fernando*, 537 P.2d 1250 (Cal. 1975); *In re Application U-2*, 413 N.W.2d 290, 298 (Neb. 1987); *W. Maricopa Combine, Inc.* v. *Ariz. Dep't of Water Res.*, 26 P.3d 1171, 1175–1176 (Ariz. Ct. App. 2001); *S.W. Sand & Gravel, Inc.* v. *Cent. Ariz. Water Conservation Dist.*, 212 P.3d 1, 5 (Ariz. Ct. App. 2008).

[73] CVWD Summary Judgment Phase 2 at 22.

[74] Ibid.

[75] CVWD Summary Judgment at 22; See *Cent. & W. Basin Replenishment Dist.* v. *S. Cal. Water Co.*, 109 Cal. App. 4th at 912.

The final argument by CVWD applies the *Kimbell Foods* test for applying state law to federal common law questions.[76] These factors determine if a "nationally uniform body of law [is] to apply in situations comparable to this, whether application of state law would frustrate federal policy or functions, and the impact a federal rule might have on existing relationships under state law."[77] Since Native American property rights are derivative of federal common law, state law property rights do not automatically apply on reservations. CVWD argued that the court should apply state law principles to pore space ownership and not craft new federal principles.[78] Applying the three-factor test, the CVWD argued that crafting a new Native American tribal right to pore space under federal common law would damage water management in the West.[79] Nonuniform rights to pore spaces in and outside reservations would lead to issues of conflicts in water management.[80]

The DWA pore spaces arguments mirrored the CVWD's arguments, but placed emphasis on the public resource aspects of the storage space within aquifers.[81] The DWA argues that the term "pore space" was inappropriate in this circumstance, and that "storage spaces" should be used instead.[82] Citing case law, the DWA argues that pore spaces are used in legal contexts to refer to mineral rights, but the term storage rights are associated with groundwater.[83] The DWA infers that the Tribe is using a *cujus est solum* doctrine discussed in *Causby*.[84] Under this doctrine, the surface owner owns all materials from the core of the earth and the atmosphere above the land. The doctrine was limited in *Causby* to allow public use of the atmosphere for commercial use by aircraft.[85] The DWA describes how the cujus est solum doctrine has been firmly rejected by courts addressing ownership of groundwater. The DWA further argues that a Tribal reservation of pore spaces would require compensation to neighboring properties where the Tribe's small aquifer replenishment program is active.[86] The land was issued to a railroad company before the Tribe's reservation was established, requiring the Tribe to compensate the company for water flowing into these parcels.[87] Critically, the

---

[76] CVWD Summary Judgment at 24; See *United States* v. *Kimbell Foods, Inc.*, 440 U.S. 715 (1979).
[77] *Wilson* v. *Omaha Indian Tribe*, 442 U.S. at 673.
[78] CVWD Summary Judgment at 24.
[79] Ibid.
[80] Ibid. at 25–26, stating, "As a result, any rule that gave an overlying Tribe or tribal owner the right to object to groundwater recharge if any of the groundwater seeped under its property could force CVWD, or anyone else wishing to recharge groundwater including the Tribe, to negotiate with scores of individuals and entities for permission. Any breakdown in negotiations would prevent valuable recharge efforts. Such a rule, moreover, would raise immense proof issues, not to mention invite endless disputes and litigation. Fogg Decl. Ex. A at 17–22."
[81] DWA Summary Judgment Phase 2 at 2.
[82] Ibid.
[83] Ibid.
[84] *United States* v. *Causby*, 328 U.S. 256, 260–261 (1946) ("*Causby*").
[85] See *Causby* at 260–261.
[86] DWA Summary Judgment Phase 2 at 12.
[87] Ibid.

approach of the DWA associates the storage right with groundwater rights, not as an independent legal argument under property law.[88] However, the Tribe made no claim to the ownership of groundwater and misstates the Tribe's *property right* claims for pore spaces.

The Tribe's claims rested on broader areas of law and not on an implicit storage right under groundwater law. The Tribe claimed that pore spaces are akin to mineral rights, where the property is held in trust by the United States for the benefit of the Tribe.[89] Under this theory, the property rights are defined under federal common law.[90] Property rights defined under federal law cannot be extinguished without federal consent.[91] Under this doctrine, no references to state law property rights are required unless the *Kimball Foods* factors are met.[92] The Tribe's arguments, however, sidestep the *Kimball* balancing test in favor of citing previous cases that have affirmatively shown that tribal property rights include "all constituent" elements of the subsurface.[93] However, the Tribe also argues that statutes and other case law have established a property right to pore spaces.[94] Within California, the courts have found trespass for the flow of wastewater into the mineral estate of others.[95] The Tribe also discusses various statutes and non-California cases that have also found a property right to pore spaces.[96] These cases involve oil, gas, and minerals, which are traditionally nonpublic resources. While the Tribe mentions these cases to support their argument, the Tribe does not argue that the rights are established under state law.[97] The Tribe argues that the constituent elements of the reservation are held in trust by the United States under federal common law.[98] The Tribe also carefully avoided the standing issues raised in the agencies' arguments by referencing the question posed in phase 2, which only addresses the potential existence, not actual determination, of pore spaces rights.[99] Strangely, the United States did not mention the arguments of pore spaces in their submissions to the court.[100]

---

[88] See Ibid. at 4.

[89] Tribe Summary Judgment Phase 2 at 19.

[90] Ibid. at 19, citing *Wilson v. Omaha Indian Tribe*, 442 U.S. 653, 670 (1979).

[91] See *Wilson v. Omaha Indian Tribe*, 442 U.S. at 671.

[92] Ibid. at 673.

[93] Tribe Summary Judgment Phase 2 at 19, citing *United States v. Shoshone Tribe of the Wind River Reservation in Wyo.*, 304 U.S. 111, 116 (1938).

[94] Tribe Summary Judgment Phase 2 at 19–20.

[95] Ibid. at 20, citing *Starrh & Starrh Cotton Growers v. Aera Energy, LLC*, 153 Cal. App. 4th 583, 592 (2007).

[96] Tribe Summary Judgment Phase 2 at 19–20, citing Wyo. Stat. Ann. § 34-1-152(a); N.D. Cent. Code § 47-31-03; Mont. Code Ann. § 82-11-180(3); *Mosser v. Denbury Res., Inc.*, 898 N.W.2d 406 (N.D. 2017); *Dep't of Transp. v. Goike*, 560 N.W.2d 365 (Mich. Ct. App. 1996); *Emeny v. United States*, 412 F.2d 1319 (Ct. Cl. 1969); *United States v. 43.42 Acres of Land*, 520 F. Supp. 1042, 1045 (W.D. La. 1981).

[97] Tribe Summary Judgment Phase 2 at 20.

[98] Ibid.

[99] Ibid.

[100] U.S. Summary Judgment Phase 2 at 6–13.

The two positions of the litigants on the use of pore spaces reflect the emerging issues surrounding the legal definition of an aquifer. On one extreme, private pore spaces would protect the Tribe's rights to store water in the aquifer, but also directly interfere with the agencies' current attempts to balance water use in the basin. A completely public pore space definition could prevent the Tribe from utilizing the pore spaces to store its own water and likely would be a violation of the Tribe's sovereignty over its reservation lands. The Tribe's arguments for a water-quantification standard echo the arguments for pore spaces, showing a preference for expansive uses of previous federal cases and no deference to state law.

8.2.4.2 Quantification Standards

While the practicably irrigable acres (PIA) standard has become the predominant method of quantification for Native American reserved surface water rights, it has not been clearly established for groundwater rights. The PIA standard is derived from the "purpose" of the reservation under the *Winters* doctrine.[101] The creation of the reservation carries with it an implied reservation of water in a quantity to fulfill that purpose. Often, that purpose for Native American tribes was the establishment of a homeland, based on agriculture.[102] Because the origins of the *Winters* doctrine occurred before the rise of artificial lift of groundwater wells, groundwater was not considered when tribes first claimed the need for water.

Additionally, groundwater may allow land that would not be irrigable otherwise, changing the area of land and complicating the calculation of the PIA. Therefore, the standard of quantification for groundwater may have other purposes, possibly including agriculture, not limited to the PIA standard. Courts apply the PIA standard by determining the number of acres potentially available for irrigation and the volume of water required for farming that land, and for the maintenance of livestock on that land.[103] Further, water volumes established by the PIA standard are not limited to agricultural uses, and may be used for commercial, domestic, or industrial uses.[104] The original purpose, often to create a homeland for a tribe, may offer alternative paths and standards for quantification.

The Tribe argued, first, that the PIA standard should apply to the reserved groundwater rights. As established by prior litigation, the purpose of the Tribe's reservation is to "establish a home and support for an agrarian society."[105] The Tribe argues that additional water was reserved for the homeland purposes on

---

[101] See *Cappaert* v. *United States*, 426 U.S. 128, 141 (1976).

[102] *Arizona* v. *California*, 373 U.S. 546, 600 (1963).

[103] See *Arizona* v. *California*, 439 U.S. 419, 421–422 (1979).

[104] See Ibid. at 422.

[105] Tribe Summary Judgment Phase 2 at 8.

lands that cannot be irrigated, adding a homeland volume to the PIA volume.[106] These two potential sources of reserved rights were the standards advocated for by the Tribe.[107] The homeland standard, however, has limited precedent and unclear specific methods of quantification.

The homeland standard is one supported by the Tribe's coplaintiff, the United States. The United States argues that the PIA standard is not applicable in all situations.[108] Instead, the United States claims that a more holistic balance of factors is required to determine the volume of water required for the reservation.[109] Drawing from case precedents, the United States suggested using nine factors: executive orders establishing the reservation, historical evidence of water use, need based on changed circumstances, cultural water uses, hydrological conditions, economic circumstances, the Tribe's present and future population, present and future water needs, and any other relevant factors.[110]

The CVWD argued that only the PIA standard should apply, but that no attempt to withdraw groundwater has occurred, preventing the court from ruling on the case.[111] Like the arguments over pore spaces, CVWD has not interfered with the Tribe's use of groundwater and state law allows access without the need to federally reserve those waters.[112] Additionally, other groundwater users in the region were not included in the litigation, and a ruling by the court would not redress the alleged groundwater injuries.[113] However, the CVWD claimed, if its standing and ripeness arguments fail, that the PIA standard should apply.[114] Since the PIA standard has not been applied to groundwater, the CVWD suggested a quantification method using the sustainable yield of the basin.[115] This would be akin to determining how much water is available in a river or groundwater basin, setting a maximum total for the basin. This argument undermines the need-based assessment under the PIA standard, preventing all the irrigable land from being used to calculate the water needs. Instead, the CVWD uses a "modified PIA" that uses the resort economy (presumably less water than would be needed for agriculture) to determine the water needed for the reservation.[116] Finally, the CVWD argued that all water available under state law should be subtracted from the volume of water determined by whatever standard is used.[117] This would

---

[106] Ibid. at 6.
[107] Ibid. at 13.
[108] U.S. Summary Judgment Phase 2 at 9.
[109] Ibid.
[110] Ibid. at 12.
[111] CVWD Summary Judgment Phase 2 at 11.
[112] Ibid. at 11–12.
[113] Ibid. at 12.
[114] Ibid. at 14.
[115] Ibid.
[116] Ibid. at 17.
[117] Ibid. at 21.

effectively eliminate any volume reserved, since California law currently allows considerable groundwater withdrawals.

The DWA argued that the homeland standard is the only appropriate one for groundwater determinations.[118] Because the Tribe no longer relies on agriculture, and mostly supports a resort industry, "changed circumstances" warrant use of the homeland standard.[119] The CVWA suggests the homeland standard takes many factors into account, including the size of the Tribe, the sustainable yield of the basins, the amount currently available under state law, current access to surface water, and land area.[120] The DWA claims that only the minimal amount of water to fulfill the reservation's purpose should be allocated under the *Winters* doctrine.[121]

8.2.4.3 Water Quality Components

The water quality aspects of the *Winters* doctrine have little precedent but appear to be a logical extension of the purpose-based reservation of water. Water reserved for a specific purpose (to create a homeland) would require that the water actually fulfill the domestic and agricultural needs of the tribe. However, beyond the general-purpose analysis of the *Winters* doctrine, no standards (like the PIA in quantification) have been developed. As there is no standard, this court will be the first to directly address this aspect of the doctrine when it decides phase 2 issues.

The Tribe follows this logic, arguing that water quality is implicit in the doctrine, since otherwise the purpose of the reservation would be entirely frustrated.[122] In similar circumstances, another tribe was able to protect reserved water from excess salt content that would prevent it from being used in agriculture.[123] Other cases cited refer to water temperature and tribal rights to fishing being degraded by inhospitable river temperatures.[124] Further, the Tribe argues that the water is property of the United States and may not be degraded without explicit consent from the government.[125] Because the United States holds a possessory interest in the water in the aquifer, it cannot be lost or damaged without authorization.[126] Further, the Tribe argues that the water quality requirements are compatible with state law, which discourages contamination of groundwater resources.[127]

---

[118] DWA Summary Judgment Phase 2 at 22.
[119] Ibid. at 24.
[120] Ibid. at 25–30.
[121] Ibid. at 32.
[122] Tribe Summary Judgment Phase 2 at 16.
[123] Ibid.
[124] Ibid. at 17.
[125] Ibid.
[126] Ibid. at 18.
[127] Ibid.

The United States also argued that *Winters* rights establish a water quality standard. The United States infers from the doctrine that contamination of groundwater effectively interferes with the Tribe's use of groundwater, just as other well users interfere under quantification questions.[128] Under the United States' argument, any effective reduction in potential water uses would be treated the same.

The CVWD argues that no water-quality protection is supported by the *Winters* line of cases. First, CVWD argues that current state law specifically addresses groundwater degradation and federal law regulates the salt content of the Colorado River water used to replenish the aquifer.[129] The CVWD suggested that the Tribe could use state-based nuisance laws to address their water-quality concerns without resorting to a new federal rule.[130] Since no cases have directly implemented a federal standard for reserved water right, there was no justification for the court to extend the doctrine to that purpose.[131] The DWA's arguments follow the same pattern as the arguments offered by the CVWD. The CVWD denied that a water-quality component is included in the doctrine, and also argued that the Tribe has not suffered an injury.[132] Since the water could still be used by the Tribe (the imported water quality meets federal and state standards) and the Tribe does not use any groundwater currently, there is no standing for the issue.[133] Further, California law prevents the kinds of injuries that the Tribe claims.[134] However, the court may not wish to find that a reserved water right may be contaminated by others to any degree, not even federal standards.

8.3 The End of the Road? The Court Grants Summary Judgments for Some Issues

On April 19, 2019, the United States District Court for the Central District of California issued a decision on the motions for summary judgments submitted by all four parties (the Tribe, USA, CVWD, and DWA) related to standing and ripeness for the issues described earlier. For the question of the quantification of the Tribe's reserved groundwater, the court determined since the Tribe did not currently pump groundwater from its reservation, the Tribe could not show an "injury in fact" and that the actions of CVWD and DWA "actually or imminently harm the Tribe's ability to use sufficient water to fulfill the purposes of the

---

[128] U.S. Summary Judgment Phase 2 at 15–16.
[129] CVWD Summary Judgment Phase 2 at 28–29.
[130] Ibid. at 29–32.
[131] Ibid. at 34.
[132] DWA Summary Judgment Phase 2 at 18–20.
[133] Ibid. at 20.
[134] Ibid. at 21.

reservation." However, the court noted that "non-use does not destroy the Tribe's federally reserved water right."

Likewise, the court decided that the Tribe did not show sufficient evidence of harm to water quality to support its claims of injury, even if TDS in groundwater will increase due to CVWD and DWA's actions. The Tribe had failed to show that the decrease in water quality would prevent the water from fulfilling the purposes of the reservation.

As to pore spaces, the court made a careful distinction between *ownership* of the pore spaces and the *right to fill* the pore spaces within the aquifer. Similarly, the court distinguished its ruling on the declaration of the Tribe's ownership interest in pore spaces from the Tribe's requested injunction preventing CVWD and DWA's infringing on that ownership. As to the declaration, the court decided that the Tribe had not provided sufficient evidence that CVWD and DWA had injured the Tribe's ability to store water in the pore spaces and did not have standing to make the claim of injunctive relief. As to the declaration of an ownership interest, the court in its decision deferred to phase 3, when the court would consider "whether the Tribe has rights to sufficient pore space to store its federally reserved water right."

Importantly, the court refused to hear arguments regarding ownership of pore spaces as a component of the surface estate, but only considered arguments connecting any ownership to the reserved water right. Using a careful analysis of the Tribe complaint, the court states in footnote 16:

The Court notes that in its Phase II briefing, the Tribe contends it owns the pore space underlying its reservation because it is a "constituent element" of the surface estate the Tribe owns. (See Dkt. Nos. 203–1 at 19–20; 217 at 2, 18; 298 at 4–5.) This directly conflicts with the Tribe's Complaint, which asserts the ownership interest is in "sufficient pore space in the Groundwater Basin aquifer underlying the Coachella Valley and the Tribe's reservation to store its federally reserved right to groundwater for all present and future purposes." (Compl. ¶¶ 66, 76.) All paragraphs of the Complaint address the pore space interest in terms of use connected to the Tribe's asserted federally reserved right – not as subterranean, constituent elements of land ownership. (See id. ¶¶ 8, 10–13, 55, 66, 71, 76.) Thus, the Tribe's argument that it owns the pore space underlying its reservation as a constituent element of its surface estate is not properly before this Court.

The Tribe's complaint rested its argument to the ownership of aquifer pore spaces within the context of its federally reserved water right, but not as an inherent part of its ownership of the surface estate. With this distinction, the court's determination will only relate to the volume of pore spaces related to the Tribe's federally reserved water right and not a generally applicable statement regarding ownership of pore spaces under the reservation.

As of this writing, it appears that the Tribe filed an amended complaint that will presumably alter the claims originally made by the Tribe related to pore space ownership. However the dispute is eventually resolved, it appears that the court will not be taking a firm stance on whether the Tribe owns the pore spaces below the Tribe's reservation independently from its federally reserved water right. As discussed later, this will not be the first court to confuse water-related rights with questions of exclusive ownership of pore spaces.

8.4 What Could Have Been and What Could Be: Our Guesses

If we peer into the crystal ball and imagine how the Agua Caliente litigation could be resolved, absent the issues with standing, complaint terminology, and injury, the litigation would need to answer a question that few courts, if any, have directly addressed. As discussed earlier, it appears that the Agua Caliente court will continue in the tradition of merging water law principles with the mainly property law question of exclusive ownership of aquifer pore spaces. In the following sections, we place ourselves in the position of judge and look past the procedural questions and dive into the substantive question: who owns the aquifer?

8.4.1 Pore Space Ownership

Applying the *Winters* doctrine to groundwater would not have been a simple task for the Agua Caliente court. It could determine a volume and a location for withdrawal, but that would oversimplify the situation. Groundwater may have a wide range of natural quality, making some groundwater useless for irrigation, drinking, or other uses. What volume of poor-quality water fulfills the federal purpose if very little meets its needs without significant treatment? Does the potential cost of treatment limit or expand the reserved volume? Groundwater quantities are not as simple as opening and closing a river diversion, but depend on pumped depth, aquifer characteristics, or hydraulic pressures. Groundwater does not have a meaningful annual pattern of flow, which would allow a court to decide on a simple cfs rate at a specific location.

Groundwater is stored diffusely across the landscape, with no particular series of well locations controlling flow. Setting a volumetric flow or a specific water-table level may oversimplify the evolving needs of the Tribe and limit, rather than promote, the future of the Tribe. May a tribe intentionally contaminate its aquifer with wastewater or hazardous waste (into poor water quality zones, for example)? Does a "homeland" purpose include disposal of waste? Because groundwater is inexorably linked with the aquifer (storage medium), any right to groundwater also indirectly means a right to aquifer resources (storage, biological purification action, distant recharge zones, hydraulic pressures, etc.). Rights to groundwater necessarily apply to the non-water portions of the aquifer area. As the Tribe argues in this case, this implicates property issues at the same time as water rights. Does the *Winters* doctrine apply to non-water resources, like pore spaces or a specific hydraulic pressure in a confined aquifer? While these are issues that will likely trouble lawyers and hydrogeologists, depending on the outcome of the case, the court will likely simplify these issues to a cfs rate determined by interpreting vague terms like "homeland purpose" or "practicable irrigable acres." With the need for a simple and satisfying remedy, the court will probably use simple answers to the complex legal and physical issues at play in the litigation.

The court would have likely ruled that the "storage spaces" of the aquifer are a public resource. The argument posed by the DWA on this issue creates a distinction between pore spaces and storage spaces: the former being used in mineral law and the latter in groundwater law. The narrow legal distinction, while probably satisfying a court's concerns with overbroad applications of the precedent, separates hydrocarbon, carbon sequestration, and other uses of pore spaces from groundwater aquifers. Unfortunately, the distinction would likely not limit the scope of the ruling. Hydrocarbon reservoirs contain water, minerals can incorporate water in their formation and structure, and there is no meaningful distinction physically between pore spaces. In addition, how contaminated must a water supply be to become a mineral? Any distinction using the current or potential storage of irrigation or drinking water is overbroad. As discussed later, a sounder basis lies in historical established public use under the law of easements. Future cases will be forced to decide how much and what kinds of water distinguish "storage" and "pore" spaces when the rocks are the same otherwise.

Quantification of the groundwater right would likely take the form of a specific annual volume determined by land area, population, and potential groundwater use and not the "homeland" standard. The court would likely not incorporate aquifer properties or the other hydrogeologic components. Because the court has only a handful of surface water cases, the court will likely wish to stay within previous precedent. While some courts have deviated from the PIA, they have stayed within the confines of a specific annual volume, not river levels, flow rates, or other seasonal hydrological properties. Because of this tendency to simplify quantification to annual volumes, the court will likely not incorporate the sustainable yield of the aquifer into the federal reservation. The court will likely determine the area of land that could be irrigated for agriculture, using wells, and reserve the needed volume to irrigate it. The rest of the aquifer users will have to work around those reserved rights, just as surface water reservations are incorporated into the prior appropriation system. Whatever remains of the sustainable yield of the aquifer (a limit and concept imposed by state law, not federal law) would not be available for other state-law-based groundwater users.

Water quality arguments would be likely to favor the Tribe, but may not be as clearly a "win" as they had hoped. The key operational language from the *Winters* doctrine is the necessary fulfillment of some "purpose." Here, the purpose is the creation of a viable home for the Tribe's members. While the Tribe surely has a right to be free from unreasonable groundwater contamination, it isn't clear that the Tribe has actually been harmed by the actions of the water districts. In fact, the Tribe has benefitted from the aquifer recharging flowing into their portion of the aquifer. While the water may be somewhat lower in quality than naturally occurring groundwater *in some locations*, these issues are overshadowed by the more impactful quantity questions. The Tribe could be easily seen as a free rider in the aquifer recharge program. In a strange twist, if the Tribe was successful with the water contamination arguments, they would effectively force the agencies to cease recharging their portion of the aquifer. With the Tribe's limited access to surface water unable to offset their own uses, the Tribe may end up worse overall in terms of water availability. After all, slightly dirtier water is better than none at all. While water of a reasonable quality is likely an implicit component of the *Winters* doctrine, it isn't likely that the Tribe has been sufficiently harmed by the recharge activities for a court to issue any remedies on this issue.

8.4.2 Policy Options for Pore Spaces: Staring into the Void

Policies for pore spaces will need to protect their current and potential uses for society. Whether that is private property or public ownership, pore spaces serve a critical role in aquifers, oil reservoirs, waste disposal, and potentially future carbon sequestration. Some kind of system must be established by the courts or by law to determine which pore spaces are private, public, or a combination of both. While a "grand unification theory" of pore spaces may not be legally necessary, more conflicts over use and control of pore spaces are likely to arise in the future. Reviewing the earlier cases, a few alternatives emerge that could guide future policy-makers and courts as they wrestle with the scope of rights to pore spaces.

8.4.2.1 By the Depth Alone

The approach that is implied by the *Causby* case sets up a spectrum with the earth's surface as the center. At the extremes of depth and height, the public has the most likely ownership. Because these extremes better serve society's public needs (travel, for example), these extremes were never a component of private property. Private property is the portion near the surface of the earth. The depth of oil and gas wells, aquifers, and skyscrapers would be the extent of private ownership.

This approach works well in the context of the atmosphere under circumstances like those in *Causby*. For the subsurface, it produces unexpected results that do not follow the outcomes of the cases listed earlier. Oil and gas reservoirs are much deeper than the typical groundwater well, showing that private property would have intervening layers that would have to fall into an exception to the rule. The exception used in the *Causby* balancing test was a public interest that outweighed the private need to exclude. For an aquifer, the balancing test could fall either way and does not fit cleanly into the expected pattern. If public uses of aquifers outweigh private rights, then the *Causby* pattern would imply that any pore spaces below the aquifer would also likely be public rights. Since this implies that oil and gas reservoirs would become public property, the court would not likely apply this kind of test to all pore spaces.

8.4.2.2 By the Original Contents

Courts have implicitly been determining the rights to pore spaces based on their original contents in the case. Under this method, the pore spaces would be divided into types based on their contents, like oil, gas, water, or other materials. The property rights associated with the material would then determine if the storage rights are public or private. Many of the cases dealing with water have followed this approach, finding that the water in the aquifer showed that the public rights associated with water also transferred to the pore spaces holding it, forever. This approach is not used with mineral rights. As shown in the cases above, rights to extract methane (a private mineral estate right) did not extend to the pore spaces, which were owned by the landowner. The material within the pore spaces did not determine the ownership of the pore spaces themselves. This approach would not comply with cases within the mineral law context.

The original content approach would also potentially limit pore space uses as technology improves and storage needs increase. As mentioned earlier, emptied pore spaces have the potential to be used for other resources. A drained fossil water aquifer could conceivably be used to store hazardous wastes. Under this approach, the public rights would be attached to the empty fossil aquifer forever, preventing its use by private parties. While seemingly easy to apply, the original content approach may not serve the public's interest in pore space utilization.

8.4.2.3 By Ad Hoc Best Utilization Method

Another approach (that may be implicitly the current approach used by courts) is an ad hoc approach. Under this method of determining the ownership of pore spaces, each case of conflict is heard and decided based on a balance of equities within that situation. A court could apply a factored test for the highest and best use of pore spaces. For example, the test could weigh the number of current pore space users, the original use of the pore spaces, and potential public needs against landowners' burdens, investment-backed expectations, and economic costs (and opportunities). The ad hoc test would likely achieve outcomes that get close to an equitable solution, but would require great time, expense, and scientific investigation to complete. The outcomes from these cases would be unpredictable and increase the risk associated with any kind of utilization of pore spaces, either by the public or by private organizations. The potential for years of research, legal conflicts, and unpredictable outcomes could raise financial risks and prevent the best use of pore spaces.

8.4.2.4 *The Possible Middle Road: Public Easement Law*

The final alternative expressed by the earlier cases involves a separation of the "sticks in the bundle" of property law. Servitudes and easements have deep roots in the common law of property. These allow the separation of the use of a certain portion of property from the absolute ownership interest in that property. Easements are commonly used for driveways, sidewalks, ditches, and many other types of use-based sharing of private property. Public navigation servitudes allow boats to travel on rivers and the construction of navigation improvements. On the Oregon Coast, a public right to use beaches has been established under a similar doctrine based on "custom." Another related doctrine, called the public trust, allows access to waterways otherwise inaccessible to non-landowners.

Applying these doctrines to pore spaces could allow the "servitude" referenced in the *Niles Sand & Gravel Co.* v. *Alameda County Water Dist.* case to inform a method of deciding what pore spaces remain private, while allowing public storage of groundwater. Many states recognize that water is owned by the public within their state constitutions, the founding document of all laws within the state. This public ownership could imply that the pore spaces holding the public's resources also have an *implied reserved easement* for storing water within aquifers at statehood up to their natural water table. This legal approach mirrors the *Winters* doctrine of federal law. Instead of reserving a portion of water from a river for federal purposes, the state is *reserving a portion of the storage space from private landowners for state purposes implied by the constitution*. A reserved public easement would cleanly divide pore spaces typically used by mineral law for oil and gas recovery from the public's uses of aquifers. Since the level of the water table at statehood would determine the upper extent of the public's easement, landowners would be protected from overfilling of pore spaces (and filling those pore spaces of others), that interferes with their own use and enjoyment of their property. The analysis used by the court would come from customs and practices used implicitly throughout the cases cited earlier.

8.5 Conclusion

The Agua Caliente litigation, while originally meant to protect the Tribe's aquifer from contamination and neighboring groundwater uses, is not so simple. The Tribe's claims to pore space ownership have the potential to open fissures between water and property law. The court deciding the case has indicated that it will follow the trend, by letting water law inform its decision on pore spaces, finding a public ownership of the aquifer. While this litigation will end, the conflicts and potential benefits of pore spaces will expand in the future.

Climate change is the new paradigm of natural resource management. Pore spaces, while providing the hydrocarbons that fuel climate change, could also provide some potential relief. Carbon sequestration and waste disposal alleviate some of the materials that harm our environment. Water storage is the emerging issue as the world braces for the effects of climate change. Aquifers have the potential to store huge quantities of water below the surface of the earth. The rights associated with those pore spaces will determine how fast, efficiently, and beneficially this resource can be adapted to help society brace for a warmer, drier climate.

9

Application to an Aquifer System

Harney Valley Area of Concern

The Harney Valley ("the Valley"), like other regions in Oregon, has aquifer problems, and a CAGA could give new life to the region's aquifers. The Valley is famous for elevated arsenic in groundwater (Smitherman, 2015). The community has already experienced groundwater declines due to pumping for agriculture, domestic, and industrial uses (House & Graves, 2016). In 2016, the Oregon Water Resources Department created the unprecedented Greater Harney Valley Groundwater Area of Concern (HVAC) to study groundwater declines, halt groundwater development, and possibly investigate the declaration of a critical groundwater area (Oregon Administrative Rule 690-512-0020). As the study of the groundwater conditions continues, the need for new aquifer governance strategies will increase. A "Harney Valley Collective Aquifer Governance Agreement" could be the potential solution to the HVAC's governance problems.

9.1 Harney Valley Transresource and Social-Legal Systems

This chapter begins with a summary of the transresource system in the Valley, describing the aquifer, groundwater, and surface water conditions in the Valley using the system analysis framework outlined earlier. Next, the social-legal system is described, providing the human system of laws, regulations, and approaches that have led to the current situation. Thereafter, the current groundwater management and governance efforts will be described, showing the need for a shift to aquifer governance in the Valley. Finally, the application of a CAGA governance system is proposed as a solution to the mismatch between the Valley's transresource and social-legal systems.

9.1.1 The Harney Valley Transresource System

The Harney Valley of Eastern Oregon is a high, dry, and vibrant place that many animals, people, and plants call home. The Harney Valley transresource system encompasses all the aspects of the aquifer, not just groundwater alone. The Valley is a 400-square-mile plain, surrounded on three sides by volcanic mountains with Malheur Lake to the south (Leonard, 1970). Precipitation in the Valley is relatively low, with the majority of precipitation coming as snow in the mountains, and snowmelt is the primary source of spring streamflows (Leonard, 1970). The Valley is a closed basin, and all the rivers and creeks in the region terminate in Harney and Malheur lakes (Leonard, 1970). These surface streams are diverted for agricultural use for crops and ranching (Leonard, 1970). Some of the smaller creeks no longer reach the lakes because the entire flow is diverted for irrigation.

The subsurface of the Valley is composed of volcanic bedrock with valley fill layers of unconsolidated clay, gravel, and alluvial fans (Leonard, 1970). The upper unconsolidated layers of the aquifer are at most 250-feet thick. The deeper layers of the valley fill and the deeper volcanic bedrock form confined aquifers. The confined aquifer provides the main source of agricultural groundwater, whose wells produce several hundred gallons per minute (Leonard, 1970). Various faults occur across the Valley and may contribute to the overall subsurface groundwater flow southward. Another potential confined aquifer may exist at around 2,000 feet below the surface, potentially yielding geothermal groundwater due to deep mixing within volcanic regions (Leonard, 1970).

Surface water quality is generally quite good in the Valley due to the primary source: melting snow (Leonard, 1970). Groundwater quantity in the Valley is highly dependent on the local geologic conditions. The primary aquifer is the volcanic-rock confined aquifer below the valley fill, creating artesian conditions in some areas of the Valley (Leonard, 1970). In the 1970s, groundwater depletions due to wells did not have a noticeable effect on the water table, likely due to the geology of the area. Pumping likely came from intercepted recharge in the mountains surrounding the Valley historically, with a small contribution from decreased discharge to Malheur Lake and other springs surrounding the Valley (Leonard, 1970). In the 1970s, groundwater levels in wells located in the Valley had little change over time from neighboring wells' use. However, more recent information indicates that ground-water levels in the Valley may be decreasing and affecting Malheur Lake (House, 2016; House & Graves, 2016). Malheur Lake could partially be fed by groundwater moving up through the unconsolidated upper aquifer from the lower volcanic aquifer (Leonard, 1970). Other sources indicate only 1 percent of Malheur Lake water comes from groundwater sources (Smitherman, 2015). Some observation wells show a rapid decline since 1995 (La Marche, 2017). Some senior surface water right holders may also be concerned with decreases in stream flow due to increased seepage from groundwater extraction (MacDougal, 2016).

As a closed basin, the primary loss of water in the entire transresource system is evapotranspiration. Natural evapotranspiration comes from plants, rivers, creeks, and lakes. Artificial evapotranspiration come from irrigation and domestic uses. Artificial evapotranspiration provides the primary economic benefit from water, surface or ground, in the Valley.

Malheur Lake is an important bird-migration habitat, making groundwater an environmental as well as economic issue. Groundwater discharges into springs and Malheur Lake support a variety of species, including juniper trees, wetland plants and animals, insects, and other wildlife (House & Graves, 2016). Recent meetings of the Greater Harney Valley Groundwater Study Advisory Committee ("the Committee"), created at the same time as the HVAC, indicate that some members believe western juniper trees may play a role in intercepting recharge in the mountains surrounding the Valley (Greater Harney Valley Groundwater Study Advisory Committee, 2017). After recent droughts and high alfalfa crop prices, many new wells were installed, converting land from range to irrigated agriculture (House, 2016).

Groundwater quality varies widely across parts of the Valley. In the edges of the Valley near the surrounding mountains, water quality is good, with minimal mineral content. However, water quality decreases significantly towards the southern portions of the Valley near Malheur Lake (Leonard, 1970). Groundwater in portions of the basin includes excessive arsenic, sulfates, chloride, nitrate, fluoride, iron, and boron depending on the location (Leonard, 1970; Smitherman, 2015). Some wells near Hines, Oregon produce hot water, with some springs and wells producing groundwater near 170 degrees Fahrenheit (Leonard, 1970). Springs also provide cold water to streams and creeks, depending on the source aquifer.

The Harney Valley transresource system includes groundwater quantity and quality and geothermal and environmental elements, as shown in Figure 9.1. The Harney Transresource System is a complex, interconnected set of resource elements linked by physical effects and environmental systems. The main biological effects on the aquifer are intercepted recharge by upland western juniper trees, some species of sagebrush, and rangeland grasses. Water quality varies across the Valley and poor water quality is mainly from natural sources of mercury, boron, and arsenic. Significant sources of warm and hot springs occur in the Valley, likely connected to a deep aquifer around 2,000 feet below the surface. A shallower volcanic aquifer provides the main source of groundwater for agricultural users, with a smaller number of wells using the alluvial valley fill aquifer. Aquifer storage is being created as groundwater levels decrease, potentially creating opportunities for the reuse of portions of the aquifer for other purposes.

9.1.2 The Harney Valley Social-Legal System

The other main system in the Valley is the social-legal system. Groundwater users in the Valley use much of their water to irrigate crops, for domestic use, and for

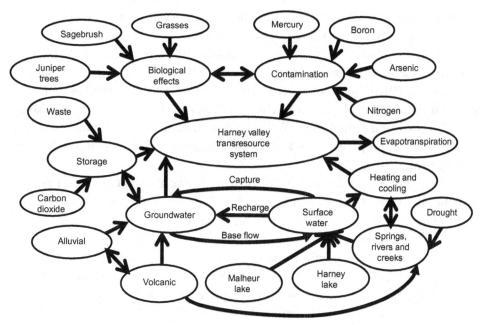

Figure 9.1 The Harney Valley transresource system

some industrial uses. Water users in the Valley appropriate both surface and groundwater (House, 2016; Leonard, 1970). Recent droughts and economic pressures have encouraged farmers and ranchers to put in new wells, transitioning historic range into irrigated alfalfa fields (House, 2016). These new and old appropriations require water rights. After noticing drops in groundwater levels, criticism by environmental groups, and some domestic users losing access to groundwater, the state placed a moratorium on any new water right applications in the HVAC while a groundwater study is completed (Preacher, 2015; Oregon Administrative Rule 690-512-0020).

Like many western states, Oregon follows the rule of prior appropriation for both surface and groundwater (Oregon Water Resources Department, 2016). The Oregon Water Resources Department (OWRD) issues water rights and administers the prior appropriation system in the state. Typically, senior water rights receive water before junior water rights (Oregon Water Resources Department, 2016). While each source is given different types of water rights, these resources are conjunctively managed (Miller, 2017). Oregon regulations presume that groundwater is hydraulically connected to surface streams if the well is within one-quarter mile of the stream for unconfined aquifers (Miller, 2017). However, wells within a confined aquifer that are within a mile of a surface stream may also be considered hydraulically connected to a surface stream if they are part of an "aquifer system" shown with substantial evidence from available hydrogeologic data (Perkowski, 2017).

Surface and groundwater rights form the major two categories of water rights in Oregon. Typically, an individual applies for a water right permit, OWRD reviews the permit for public interest, water availability, and potential connections to surface streams (for groundwater) (Miller, 2017; Oregon Water Resources Department, 2016). Instream water rights may also be held by state agencies for environmental and ecological uses (Oregon Water Resources Department, 2016).

Basin programs allow the OWRD to designate certain uses and close the basin to new appropriations in less severe situations (Oregon Water Resources Department, 2016).

Water rights allow a certain volume of water per year, at a determined rate, and specific to a certain season of use (Getches, Zellmer, & Amos, 2015). A water right's place of use, point of appropriation, and type of use may be permanently or temporarily transferred to a new location (Oregon Water Resources Department, 2016). Conditions may be placed on the transfer to prevent injury to other water users (Bonini, 2018; Koda, 2007; Oregon Water Resources Department, 2016). The OWRD does not have a systematic approach to determining injury, often using subjective opinions to determine the injury suffered and necessary conditions for transfer (Bonini, 2018; Koda, 2007).

Artificial storage and recovery projects may be conducted by large municipal entities or on a small-scale by individuals. Oregon Revised Statute (ORS) 537.534 allows a limited license to be issued for testing the viability of storing water within an aquifer, a process typically conducted by municipal bodies. Some small ASR projects have been conducted for private use, using spring-sourced domestic water and rainwater collection (Embleton, 2012; Robinson, Jarvis, & Tullos, 2017). The small-scale approach to ASR is regulated under the Oregon Department of Environmental Quality and approved by the OWRD by rule (Robinson et al., 2017).

Adaptive management in Oregon is primarily through the earlier-referenced basin programs, the unprecedented HVAC, and critical groundwater areas as shown in Figure 9.2. When a groundwater source suffers from a severe long-term decline, overdraft, or shortage, OWRD may declare the area a critical groundwater area (Miller, 2017); these critical areas restrict the kinds of groundwater uses available. The OWRD may restrict current uses, deny any new appropriations, and set preferences between types of uses (Oregon Water Resources Department, 2016). Critical groundwater area designations are often controversial, harming the local economy, and reducing the value of property held by the community (House & Graves, 2016; Miller, 2017). These designations eliminate or restrict established groundwater rights, prompting litigation and controversy. They also may reach domestic use of groundwater, implicating human rights issues (Schroeder, Ure, & Liljefelt, 2011).

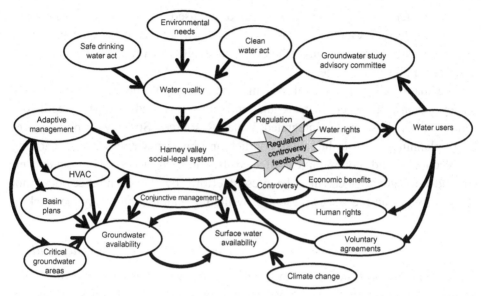

Figure 9.2 The Harney Valley social-legal system

Water is considered public property by the Oregon Constitution, which supports the use of regulations like critical groundwater areas and the HVAC to control the use of water rights in the state (Miller, 2017). The use of these top-down methods has been controversial, since they can harm the communities and the economies of affected areas (House & Graves, 2016). The regulation-controversy feedback in Oregon, as shown in Figure 9.2, has slowed the effectiveness of regulatory approaches.

The OWRD has struggled to study aquifers within the state due to limited staff, budgets, and litigation (House & Graves, 2016; Perkowski, 2018). Legislators have used budgets and their positions to prevent reductions in water consumption (House & Graves, 2016). Funding for studies of aquifers has been limited due to budget constraints and political resistance (House & Graves, 2016). Litigation opposing OWRD has also soaked up millions of dollars, preventing OWRD from hiring staff, and requiring the agency to request emergency funding (Perkowski, 2018). With limited tools and heavy resistance, the regulatory approach has had limited success in addressing the issues present in the Harney Valley, which is suffering from the regulatory-controversy feedback seen in Figure 9.2.

Oregon also has a unique approach that could break through its regulation-controversy feedback: voluntary agreements, as described by ORS 537.745. These agreements can replace the enforcement of water rights by OWRD after their approval by the Water Resources Commission (Bonini, 2018). The statute could provide for a complete reassignment of the legal definitions, conditions, and rules for water rights. These agreements could be used to bypass the regulation-controversy feedback and give considerable authority to members of the agreement to redefine groundwater rights. However, the statute has yet to be applied to any basin or aquifer in the state.

Like many other social-legal systems, the primary bottleneck of the Harney Valley social-legal system is the regulation-controversy feedback. Most regulatory actions, even those using Integrated Water Resources Strategy–inspired place-based planning, result in litigation, conflict, and involuntary restrictions, harming the economy and communities affected by reduced access to groundwater. Groundwater governance, like the present efforts by the Committee, does not completely negate the issues presented by the regulation-controversy feedback.

9.1.3 Current Groundwater Management and Governance in Harney Valley

The current approach in the Valley combines both groundwater management and groundwater governance. Groundwater management is the traditional approach to preventing resource overuse and negative externalities. In the Valley, the OWRD has used this approach during the administration of water rights and the review procedures during groundwater right applications, determining if the water use is in public interest, if water is available in the basin program, and if the water use would cause injury or surface-water interference. Further, as discussed earlier, the regulatory declaration of the HVAC, basin programs, and the potential future declaration of a critical groundwater area form the primary adaptive management components of groundwater management in the Valley. As these are primarily top-down approaches, these interventions represent groundwater management.

Groundwater governance has been achieved in the Valley. While groundwater users are not directly involved in the decision-making process, they are being incorporated into the process of studying and recommending actions to the OWRD. This participation is a moderate level of groundwater governance, allowing agencies to consult and collaborate with, but not empower, Valley residents with decision-making power (Nyerges et al., 2006). The HVAC created the Committee to gather information about the Valley's aquifer and build trust with the community (Greater Harney Valley Groundwater Study Advisory Committee, 2018). The Committee is composed of landowners, tribal representatives, state and federal agency staff, well drillers, and environmental organizations. OWRD is required by the HVAC to coordinate with the Committee in gathering hydrologic data to support a final groundwater study report, provided to the Committee for comments before finalization by OWRD (OAR 690-512-0020 (11)). This information may be used for adjustments to the basin program, or potentially for designation of a critical groundwater area.

While groundwater governance in the Valley has achieved the participation of local individuals, organizations, and tribes, the power to make decisions remains with the OWRD. The role of the Committee is limited to commentary and coordination, giving OWRD an effective veto to any suggestions by the Committee. The governance approach, therefore, results in an influence, but not control, over the regulations imposed on the Valley's community. As shown in Figure 9.2, the role of the Committee could likely result in the return to the regulation-controversy feedback and effectively a return to solutions within the groundwater management paradigm. To move to aquifer governance in the Valley, a potential Harney Valley CAGA could empower the community to move past groundwater and govern the aquifer.

9.1.4 Potential Harney Valley Collective Aquifer Governance Agreement

The potential use of a CAGA in the Valley would create a new system connecting the Harney Valley transresource system to the social-legal system, potentially solving the negative emergent properties of both. The current emergent properties of the Harney Valley transresource system are groundwater table drops, surface water depletions, and increased evapotranspiration. These are the primary negative externalities, expressed as emergent properties of the system. The emergent properties of the Harney Valley social-legal system are sources of conflict (the regulation-controversy feedback), failure to internalize negative externalities within the system of rights, and incentives to resist any regulation due to the economic harms created by reducing access to water. The Harney Valley CAGA must be able to address each of these negative emergent properties, creating a new system where positive externalities like aquifer recharge, environmental benefits, and community economic stability result. The following sections describe the process of developing and the components of a potential Harney Valley CAGA.

9.1.5 Harney Valley Collective Aquifer Governance Agreement Timeline

As described earlier, CAGAs operate in four general phases: exploration and discovery, determination of rights and shares, operation of the CAGA, and potentially abandonment or repurposing of the aquifer. The following sections expand on these ideas and apply the concepts to the specific conditions in the Valley.

9.1.5.1 Phase 1: Exploration and Discovery

Leonard (1970) provides a detailed and expansive description of the various aquifers and their properties located in the Valley. As discussed earlier, two primary aquifers and a series of rivers make up the Valley's transresources. A potential third aquifer may be the source of geothermal groundwater and the hot springs in the region. Part of the first phase of the CAGA development would be a quantification of the various aquifers' storage, artificial recharge, stream depletions, total evapotranspiration, and other relevant estimates of the basin's water mass balance. Similar to the Committee's work on a water budget, this phase would determine the physical context of the aquifers and partially support the allocation of shares in the next phase. In short, this phase determines the components and influences of the Harney Valley transresource system, presented in Figure 9.1.

9.1.5.2 *Phase 2: Determination of Rights, Shares*

The next phase of the development of the CAGA is gathering the various types of water rights, instream rights, water quality data, regulations, and laws and viewing them as components of the social-legal system. Considering the goals of the Valley and limitations imposed by law, these components are converted into shares (discussed later) that will form the basis for the CAGA voting, market, and distribution of costs and benefits. These shares attempt to incorporate the externalities (identified by law and physical impacts) and internalize them in the Harney Valley CAGA. In short, this phase adopts the social-legal system into the CAGA governance system.

9.1.5.3 *Phase 3: Operation of the Harney Valley CAGA*

In the third phase of the Harney Valley CAGA, the unit operator is selected by the shareholders and then develops the unit plan for the Valley. The unit plan is the detailed expression of the purposes and goals of the CAGA. For example, the Harney Valley CAGA may wish to avoid a designation of a critical groundwater area, and the unit plan may focus on implementing water use efficiency and abstraction taxes to curb groundwater consumption. The unit plan may also implement a mass balance-based water market and banking scheme. The third phase of the timeline links the two major domains of the agreement: the resources and the social components.

9.1.5.4 *Phase 4: Abandonment and Repurposing*

Because Oregon law does not designate specific aquifers as "fossil" aquifers with no effective level of sustainable use, the planned depletion of an aquifer and its repurposing is unlikely to be an option for the Valley. Further, the Valley community has expressed its wish to continue to live, work, and depend on the aquifer into the future (House, 2016). Components of the aquifer could be repurposed, expanding the potential of the aquifer in the long term. The deeper geothermal aquifer may serve as a new water source at first but be repurposed later for geothermal energy or waste storage. Profits from geothermal energy produced by the aquifer could be shared collectively on the basis of each geothermal shareholder's contribution to the unit and used to fund further aquifer development projects. Repurposing the aquifer adds aspects of adaptive management into the Harney Valley CAGA governance system.

9.2 Harney Valley Collective Aquifer Governance Agreement Components

The specific components of a Harney Valley CAGA will determine every aspect of the governance system for the aquifers in the Valley. This section provides the explanatory details and recommendations for a model agreement and refers to portions of the Model Collective Aquifer Governance Agreement ("Model CAGA") located in Appendix. The conceptual suggestions for the Harney Valley CAGA connect the issues, problems, and institutional aspects of the agreement to the conditions in the Valley. As a negotiated document, the following suggestions provide a potential path for the Valley, but specific circumstances in a real application of the CAGA governance system would likely vary from the suggestions below. However, many of the broad components and circumstances would likely be present in any real Harney Valley CAGA.

9.2.1 Purposes and Goals

The purposes and goals of the Harvey Valley CAGA would determine many of the other components of the agreement. For the Valley, maintaining access to groundwater would likely be the primary concern, based on the economic importance of water in the area (House, 2016). Other purposes, like preservation of the environment and restoration of Malheur Lake levels, could be incorporated into the purposes stated in the CAGA. The Harney Valley CAGA could include avoiding a critical groundwater area designation, repressurization of the confined aquifer, or development of geothermal resources in the potential deep volcanic aquifer. In the Model CAGA, the purpose is presented in Section 1 and the governance guidelines are located in Section 13. These focus the unit operator's work and the eventual unit plan.

9.2.2 Formation

The Harney Valley CAGA could be formed using three general methods: a contract, a water district, or a voluntary agreement under ORS 537.745. The Model CAGA located in Appendix (Section 1) assumes the first method. The contract approach would most reflect the type of agreement used in the oil and gas industry, whereby the only legal relationship between the parties is contractual obligations.

The second path could be the formation of a water district in addition to a contractual relationship, allowing some recalcitrant surface-water users to be forced into the CAGA using ORS 545.025 by election or county court order. A contractual relationship would still be required, despite the formation of the district, since Oregon law only recognizes irrigation districts related to surface water, not groundwater. The primary benefit would be a reduction in temporary transfers filed with the OWRD, since water may be transferred within a district without a regulatory review (ORS 540.270). The third option, the voluntary agreement, could allow the OWRD to completely forego its regulatory review procedures, or substantially change them, if the Harney Valley CAGA is approved by the Oregon Water Resources Commission. While unlikely to be approved, the third option would give a free hand to the unit operator to conduct groundwater governance within the aquifer.

9.2.3 Aquifer Unit Operator

The aquifer unit operator for the Harney Valley CAGA could be a variety of organizations, including nonprofit groups, engineering firms, law firms, or a partnership of multiple organizations. The unit operator is responsible for developing the unit plan, which would require skill with hydrological science, legal knowledge of water and property law, and charisma, as the external face of the unit and primary coordinator of aquifer projects. As the first CAGA, the Valley would establish the theoretical foundation of the CAGA governance approach, including the types of unit operators that are capable of leading a CAGA. The potential unit operator would have a limited time to serve as the unit operator, allowing the aquifer shareholders to replace them should issues arise with their leadership, as described in the Model CAGA in Section 4.

The unit operator would regulate a water market, if selected as a component of the Harney Valley CAGA. The unit operator would determine the transferability of the groundwater rights, conduct the regulatory process for the two members, and determine the hydrologic implications of the transfer for the unit. Unlike the market discussed by Young (2015), the hydrologic components would be directly incorporated into the sale and actual transfer of water allocations within the unit plan, studied by the unit operator.

9.2.4 Aquifer Unit Committee

The aquifer unit committee for the Harney Valley CAGA would be composed of all the groups and individuals in the agreement that hold shares. The committee is the final authority for major decisions in the CAGA, selecting the unit operator and the goals, and approving the unit plan. The committee members are also the primary beneficiary of the CAGA, meaning that all interested parties to the aquifer should be encouraged to join the agreement. These could include surface-water users, groundwater users, environmental groups (representing environmental water needs), government agencies, tribes, and any other party holding access to aquifer transresources. Since transresources are not limited to water, typical real estate owners with no water rights could potentially be a party to the agreement, as these parties hold property interests in pore space and geothermal rights. A conceptual description of the unit committee is located in Section 1 of the Model CAGA in the Appendix.

9.3 Redetermination

One of the most important aspects of the Harney Valley CAGA governance system is redetermination, discussed further in Chapter 5. Because of the limited data available and the complexity of aquifer systems, the initial distribution of shares would likely need periodic revision to conform to the newly acquired hydrological data. Redetermination occurs regularly in the CAGA governance system, as shown in the Model CAGA in Section 16. Shares are also redetermined if the unit boundary, discussed later, is adjusted or expanded, perhaps because new information shows that the initial description of the aquifer system was incorrect, as shown in Section 9 of the Model CAGA.

9.3.1 Information and Adaptive Management

Information is a particularly prominent issue in the Valley. With OWRD's limited budget to study aquifers in Oregon and litigation consuming its financial reserves, accurate and continued monitoring of the Valley's aquifers is lacking (House, 2016; House & Graves, 2016). Groundwater monitoring regulations have stalled due to budgetary issues and lack of legislative interest (House & Graves, 2016). OWRD has issued groundwater right permits based on limited information and often opts not to include cumulative impacts in its procedures to determine surface-water interference (House, 2016). A major component of the Harney Valley CAGA will be gathering aquifer data on the best way to minimize negative externalities and maximize the benefits (economic and environmental) of groundwater use in the Valley.

As new data is gathered and incorporated into the scientific model of the aquifers, the unit plan would be adjusted (as shown in Section 6(G) of the Model CAGA), the shares would be redetermined, and the unit boundary would be varied to match the best information available to the unit operator. With the limited budget and issues related to regulatory approaches (HVAC, Critical Areas, Basin Programs), adaptive management would be integral to the Harney Valley CAGA, not a response after impacts have become severe or excessive.

Another important source of information could be the participants themselves. Because of the large number of participants in the agreement, the need for timely and accurate information, and the difficulty of gathering all members in a single location, web-based applications could be used for voting, data gathering, and opinion polling within the unit. This soft data could be used in the development of the unit plan and as a medium for the potential water market, if implemented.

9.3.2 Unit Boundary

The unit boundary is a flexible description of the approximate location of the aquifer system. The unit boundary would include all the surface property above the various aquifers relevant to the Harney Valley CAGA, including both the alluvial, confined, and potential deeper geothermal aquifer. The primary purpose of the unit boundary is the identification of relevant individuals and organizations that could participate in the Harney Valley CAGA, based on scientific information about the aquifer system. The aquifer system would include surface-water users within the boundary, since their use may contribute to recharge of the aquifer.

As there are multiple separate aquifers in the Valley, multiple subunits with distinct share systems could create "sub-boundaries" within the CAGA. To encourage socially beneficial use of the total aquifer system, the unit operator would determine the transferability of shares between the aquifers and manage any transfer-permitting requirements with OWRD.

9.3.3 Unitized Substances and Shares

Another important component of the Harney Valley CAGA would be the determination of shares, unitized substances, and how these can be formed to produce the best governance system to meet the goals of the agreement. The Model CAGA presents various options for determining the basis for contractual shares within the agreement in Section 10. The Valley's main issues are groundwater depletion, surface water interference, and water quality. Another potential aspect of the aquifer could be the geothermal resources in parts of the Valley. Each of these components could form the basis for shares in the Harney Valley CAGA.

Currently, flows in some rivers have decreased due to alleged groundwater withdrawals, inducing recharge or decreasing baseflows in those rivers. Surface water rights could be converted into contractual shares of evapotranspiration, groundwater recharge, and return flows. Groundwater shares could be composed of classes of shares like storage depletion, stream depletion, artificial recharge, and evapotranspiration.

As a closed basin, the primary loss of water in the Valley is due to evapotranspiration, making it the ideal target for a use charge, water tax, or other form of financial disincentive based on the estimated portion of the water user's consumption. Water user "efficiency" may not be accomplished by a broadly defined water diversion charge or tax, since some of that water remains within the aquifer system as recharge. Evapotranspiration, not all water use, is also the primary loss of water in the Harney Valley aquifer system. The evapotranspiration shares reflect the externality of nonnatural increases in evapotranspiration, internalizing the negative externality of total aquifer system losses in the closed basin.

The unit operator would determine the transferability and distribution of the class of shares to individual water users based on their own use, source, and efficiency. This system could discourage socially inefficient evapotranspiration, incentivize surface-water users to allow aquifer recharge, and potentially allow a water market to form around science-based shares in aquifer transresources.

10

Getting Around Agreeing to Disagree

We have conquered Mother Nature; now we have only to conquer human nature.

—*D. R. Knowlton (1939)*

The Pretashkent Aquifer is located within the borders of two countries – the Republic of Kazakhstan and the Republic of Uzbekistan. The Pretashkent Aquifer represents the artesian basin, the structure of which includes several aquifers and complexes separating them by lower permeability sedimentary units. The area is divided into three zones: a mountain zone, foothills, and valleys. Groundwater levels are dropping due to intensive use for agriculture. The water stored in the aquifer can be considered nonrenewable groundwater due to negligible recent recharge. The water quality is apparently degrading from the high quality historically used for mineral water and medicinal consumption.

According to Puri and Villholth (2017), there is no bilateral Kazakh–Uzbek institution in place for the Pretashkent Aquifer. Kokimova (2019) indicates that the most recent groundwater storage revaluation of the Pretashkent Aquifer was conducted in 1982–1983 by the Ministry of Geology of the USSR. Safe yield was divided between Kazakhstan and Uzbekistan, at 1,464 cubic meters per day and 2,044 cubic meters per day, respectively. The two countries applied the safe yield limits until 1991 – the year the dissolution of the USSR occurred. Mechanisms of pumping controls and related enforcement have apparently been less stringent since the two countries gained independence.

There are no bi- or multilateral transboundary institutions in existence with a mandate for transboundary aquifers in Central Asia. The ongoing Governance of Groundwater Resources in Transboundary Aquifers (GGRETA) project that is being jointly implemented by the United Nations Educational, Scientific and Cultural Organization (UNESCO) and the UN International Groundwater Assessment Centre (IGRAC) is designed to build technical, scientific, legal, institutional, and hydro-diplomacy skills that facilitate collaborative management of transboundary aquifers. However, their preliminary assessments indicate that both countries appear to consolidate principal responsibilities for domestic groundwater-relevant legislation under central governmental institutional arrangements. Given this situation, regional groundwater governance does not appear to have developed dispute resolution mechanisms. In our opinion, the governance of the Pretashkent Aquifer appears ripe to apply unitization principles through a CAGA.

As part of the GGRETA project, a series of workshops were hosted by UNESCO and IGRAC. We were involved in workshops focusing on groundwater conflict management in Central Asia. We introduced the scientific mediation process, as developed by Moore Jarvis, and Wentworth (2015), where the goal was to introduce the notion of attempting to reach agreement on the merits of the disagreement over groundwater science and engineering associated with the Pretashkent Aquifer, as opposed to having personal and political biases cloud the scientific process. As part of the GGRETA training project, we suggested to the hydrogeologists and engineers in the audience that there is value in using the scientific mediation process for groundwater situations, since it moves beyond the tired and overused cliché of agreeing to disagree used by entrenched expert egos.

It seems silly that groundwater professionals cannot get along, but groundwater scientists and engineers are like other people with personal and political biases. Likewise, conflicting conceptual hydrogeologic models are also part of the formal training of hydrogeologists, focusing on the intellectual method of multiple working hypotheses introduced in the late 1890s by US hydrogeologist, Thomas Chamberlain. Multiple working hypotheses revolve around the notion of developing several hypotheses to explain observed phenomena (Chamberlin, 1897). As a consequence, groundwater professionals also have a strong personal affinity and identity to their work, given that imagination and creativity are key parts of developing their working hypotheses.

We have experimented with agreements that get around the notion of "agreeing to disagree" when it comes to working in watersheds along the coast of Oregon (Walker, 2011). We found that a "Collaboration Compact" worked well in getting stakeholders on the same page with respect to moving forward, with securing shared "science," data, and decision-making before implementation. Given the success of the Collaboration Compact concept in a complex setting such as a coastal watershed, with many different private and public enterprises, we saw value in integrating the concept into the unitization process for a CAGA, as depicted in Table 10.1.

We took the concept of developing the Collaboration Compact for the Pretashkent Aquifer one step further by integrating the four main elements of

Table 10.1. *Example of principled collaboration compact*

Stage of oil unit or aquifer unit life	Typical unitization status for oil and gas by Worthington (2011)	Proposed aquifer unitization
Collaboration		Principled collaboration compact
Exploration		Pre-unit agreement based on voluntary geographic unit
Discovery		
Appraisal	Pre-unit agreement	Pre-unit agreement based on voluntary geologic unit
Commerciality		
Development	Unitization and operating agreement	Unitization and operating agreement
Production	Redetermination(s)	Redetermination and compulsory/conservation units
Abandonment		Redetermination(s) of new use of aquifer

principled negotiation described in *Getting to Yes*, by Fisher, Ury, and Patton (2011), resulting in the example Collaboration Compact that follows:

1. Separate the people from the problem. The goal is not to "win" but to reach a better understanding of each party's concerns.
2. Focus on interests, not positions. Look beyond such hard-and-fast positions to try to identify underlying interests – their basic needs, wants, and motivations.
3. Invent options for mutual gain. In principled negotiation, negotiators devote significant time to brainstorming a wide range of possible options before choosing the best one.
4. Insist on using objective criteria. In principled negotiation, negotiators rely on objective criteria – a fair, independent standard – to settle their differences. Importantly, parties should agree in advance about which objective criteria to consult, and agree to abide by the outcome.

10.1 Collaboration Compact Example

Shared aquifer members recognize the importance of procedural clarity and commitment. This "Commitment to Collaboration Compact" articulates that clarity and investment.

The compact consists of four sections: (1) tenets, (2) context, (3) principles, and a (4) preliminary list of topics.

10.1.1 Tenets

As a "compact," this document outlines the commitment the two countries make to one another. The compact is designed to define the practice of "collaboration." The basic tenets of this compact are as follows:

One. The compact is a "living document," created, maintained, and modified by the two countries. The two countries can "revisit" the compact whenever appropriate and improve its content.

Two. The compact is a procedural guide, describing the operation of collaborative work and interaction.

Three. The participants sign the document, thereby indicating their commitment to working as a collaborative group on transboundary aquifer matters. New participants can add their signatures, and current participants can withdraw their support.

Four. The compact is an informal agreement among participants and does not have any legal standing. It does articulate a "good faith commitment" among members but does not provide any basis for legal action involving any of the compact signatories.

Five. The compact is a public document, available for anyone to read and review.

10.1.2 Context

Participants recognize the following contextual factors.

- We need to consider future generations. We should address the long term as well as the short term.
- We need to understand and respect the "divides": history, economy, environment, and world views.
- We recognize that there will be disagreements about the "facts." We need to consider the sources of and our biases.
- We acknowledge that there may be some objections and warnings. We should be aware of who is affected by what we endorse.
- We respect skepticism and welcome participation and discussion.
- Our dominant concern emphasizes the quality of life in the participating countries.
- We need to understand the implications of exploiting transboundary groundwater.

10.1.3 Principles

The participating members recognize the following principles.

- Activities described under this compact should be beneficial to both countries.
- Aquifers to be jointly studied, and the scope of the studies or activities to be done on each aquifer should be agreed upon with a mutually agreed upon framework.

- The activities should respect the legal framework and jurisdictional requirements of each country.
- No provisions set forth in this compact will limit what either country can do independently in its own territory.
- Nothing in this compact may contravene what has been stipulated in any previous treaties between the two countries.
- The information generated from these projects is solely for the purpose of expanding knowledge of the aquifers and should not be used by one country to require that the other country modify its water management and use.

10.1.4 Preliminary List of Topics

Since the time of our training in concert with the GGRETA project, one of the outcomes of the ongoing work by IGRAC in Central Asia was the development of a groundwater model to "start the conversation." According to Siepman (2019), a graduate student was recruited by IHE Delft Institute for Water Education to summarize model input preparation and data harmonization using the datasets and information from the Committee of geology and hydrogeology of Kazakhstan and results of the GGRETA project.

We integrated this information into the Collaboration Compact introduced at the training to show how Collaboration Compacts become living documents and an opportunity to prevent misunderstanding regarding the technical approaches to determinations and redeterminations as a means to prevent disputes.

Conceptual Model: The conceptual model of the transboundary aquifer will be respectful of multiple working hypotheses and will incorporate knowledge from technical data and interpretations, traditional knowledge, personal knowledge and experience, and political realities. Kokimova (2019) simplified the groundwater system into six layers.

Quantitizing Conceptual Model: Attributing numerical values to the shared aquifer system will identify sources of data used, data gaps, and boundary conditions (low permeability boundaries, hydraulic connections to surface water systems, leakage, etc.). Kokimova (2019) converted the conceptual model into a steady-state numerical model.

Storage versus Recovery: Calculations of volumes of water stored in the aquifer system will acknowledge that not all of this water can be recovered by wells or drainage mechanisms. Kokimova (2019) indicated that the last groundwater storage revaluation was conducted in 1982–1983, where the safe yield was divided between Kazakhstan and Uzbekistan at 1,464 and 2,044 cubic meters per day, respectively.

10.2 Depletion Scenarios

Planned depletion scenario: The management goal is the orderly utilization of aquifer reserves of a system with little preexisting development.

Unplanned depletion scenario or "rationalization scenario" is defined where the management goal is

hydraulic stabilization (or exceptionally recovery) of the aquifer, or

more orderly utilization of aquifer reserves, minimizing quality deterioration, maximizing groundwater productivity, and promoting social transition to a less water-dependent economy.

Value of Water is not limited strictly to economic value, and includes spiritual value and value to nature.

10.3 Mediated Modeling?

Type of Model

- Groundwater versus integrated
- In-house development
- Vendor-developed, in-house run
- Vendor-developed, vendor-run

Kokimova (2019) selected the Groundwater Modeling System software (GMS).

Calibration of Model Methodology

Kokimova (2019) reports that model calibration was not possible due to the lack of data from observation wells.

Verification and Validation of Model

- Index wells
- Spring discharge
- Water quality

Kokimova (2019) identified many data limitations with complicated verification and validation of her model. As a consequence, future collaborative data-collection efforts could focus on the following items:

- The main sources used for data collection are dated by 1971, 2010, and 2016.
- No recharge estimations and discharge measurements were obtained.
- There were no publicly available river and reservoir water levels and cross-sections.
- The information on the depth of the layers was collected from cross-sections, and the number of wells was limited, not representing the whole perimeter of the aquifer.

- Values for hydrogeological parameters, such as hydraulic conductivity and recharge rates were roughly assumed based on geology and topography, as a consequence of data availability.
- Historical data from abstraction wells were available with inconsistent time series from 1954 to 2009. Only nine wells had water level observation from 1996 until 2009. The primary dataset covered the period from 1969 until 1985.
- Previous transient groundwater flow models provided the values of the water flow budget without any units.

Post-Model Audit

According to Kokimova (2019), there is no present transboundary legal and institutional framework indicating data and information sharing in the Pretashkent Aquifer. The main sources used for data collection are dated by 1971, 2010, and 2016.

Redetermination

- Groundwater circulation in the Pretashkent Aquifer is transboundary. A few wells in the territory of Kazakhstan likely pump groundwater from Uzbekistan.
- The future development plan on the Pretashkent Aquifer might be established after simulating the transient model with current abstraction rates.
- There needs to be a mutual agreement on limits on abstraction rates between the aquifer sharing states of Kazakhstan and Uzbekistan (Kokimova, 2019).

Abandonment and Reuse of Aquifer

The abandonment and reuse of the aquifer could be addressed in this section, if desired.

Dispute Resolution

The Interstate Commission for Water Coordination (ICWC) is already established in the Syr Darya River Basin with the role to organize international water management including groundwater in Syr Darya River Basin (Kokimova, 2019).

11

Serious Gaming and Unitization

Scientific mediators attempt to tread the path between Merchants of Doom and Merchants of Doubt as Merchants of Discourse using multiple working hypotheses and multiple ways of knowing as their moral compass.

—*C. Moore et al. (2015)*

In our work in water rights, water law, and water conflict resolution, we regularly use games to complete test runs of various scenarios. Serious games are useful because they provide a structured environment in which learning, research, and joint fact finding can occur – and they are fun. We have inventoried over forty games related to water in our research, and a comparable number with minor overlap with our inventory was reported by Aubert, Bauer, and Lienert (2018). We also found a dearth of academic publications or popular media regarding the various games; like Aubert et al. (2018) we often discovered the games anecdotally. Rather than delve into the knotty details of game theory related to water resources, or the rich literature on social learning associated with water games, we refer the interested reader to Aubert et al. (2018). The following synopsis of gaming groundwater and aquifer situations is adapted from Jarvis (2014, 2018a,b).

11.1 Gamifying Aquifers: How Resources and People Matter

Games focusing on groundwater and aquifers come in many forms, ranging from role plays, board games, computer-assisted board games, and online games. Games dedicated to groundwater situations span development over the past thirty years (see Table 11.1). In his blog, *The Consensus Building Approach*, Susskind (2012) writes, "There are various ways games can be used to inform, and even alter, high-stakes policy negotiations ... but this only works when the actual negotiators take part in the game in advance of undertaking their own 'real life' interactions."

Table 11.1. *List of serious groundwater games*

Game	Situation	Year	Developed By
Water on the West Bank Role Play	Water well siting, aquifer depletion	1988	Harvard Program on Negotiation
Managing Groundwater beneath the Pablo-Burford Border Role Play	Agriculture water quantity and quality across borders	1996	Harvard Program on Negotiation
Santiago Board Game	Diversion of spring water to canals for plantations	2003	AMIGO Spiel
International Groundwater Negotiation Role Play	Hydraulic connection to a transboundary water resources	2007	FAO training manual for international watercourses/ river basins including law, negotiation, conflict resolution, and simulation training exercises
Tragedy of the Groundwater Commons Computer Assisted Role Play	Hydrogeologic capture analysis and economics of pumping wells between many well owners and a lake	2013	International Groundwater Assessment Centre (IGRAC)
Groundwater Protection Dueling Expert Role Play	Wellhead protection and aquifer protection boundaries across the urban/rural divide	2014	Contesting Hidden Waters: Conflict Resolution for Groundwater and Aquifers
California Water Crisis Board Game	Groundwater use and depletion across three regions for agriculture, ecosystems, and urban growth	2014	Firstcultural Games
The Edwards Aquifer Case Role Play	Groundwater, common law rule of capture, Endangered Species Act, and role of science	2016	The Program for the Advancement of Research on Conflict and Collaboration (PARCC), Maxwell School of Citizenship and Public Affairs, Syracuse University
Save the Water Board Game. Online Version Available	Agriculturists struggle with profitable cropping and groundwater depletion	2017	Zurich University of the Arts (ZHdK) with ETH Zurich
The Groundwater Game Computer Assisted Board Game	Water users and managers play with the California Sustainable Groundwater Management Act	2019	Environmental Defense Fund and the University of Michigan
The Unitization Game	Simulation of the negotiation and dynamics of developing an aquifer unitization agreement	2020	This book

Scientific mediation is used by groundwater scientists and engineers as part of broader impacts to the general public in matters where the technical jargon and high levels of uncertainty lead to a stalemate on decision-making. Scientific mediation is also used to resolve disputes between groundwater scientists and engineers who live and work across boundaries separating many different scales, ranging from the urban–rural divide, county to county, state to state, province to province, and international.

The scientific mediation framework depicted in Figure 11.1 attempts to reach agreement on the merits of the disagreement, as opposed to having personal and political biases cloud the scientific process. While scientific mediation is a process that sounds rather utopian, it is garnering much interest by conflict resolution pracademics because it moves beyond the outdated "agree to disagree" palliative.

Serious games introduce the different types of negotiation styles, even in situations where language or cultural barriers exist. Many countries are just beginning the organization of alternative dispute resolution systems; computer-based and online games enhance their online competency in groundwater and

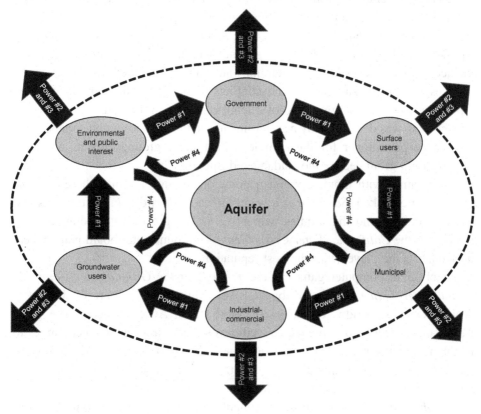

Figure 11.1 Scientific mediation framework

aquifer negotiations. Collaborative modeling is a form of serious game playing with participants developing various groundwater management scenarios. Serious gaming could also assist parties in developing Collaboration Compacts discussed in previous sections of this book.

What are the best approaches to negotiations regarding groundwater and related aquifers? In addition to the use of "search conferences" and "joint fact finding" described by Moore et al. (2015), serious games are a useful approach to addressing the "groan zone" that groundwater conflicts and negotiations enter, regardless of the scale of the conflict. Serious games can be an interactive, realistic virtual environment in which players attempt to simultaneously "juggle" growing food, growing cities, sustaining the environment, and making a profit.

One of the tried-and-true approaches to negotiation training is that of role play. Nearly every academic or professional training program in water negotiations uses role plays. Most focus on surface water allocations, water rights, benefit sharing, how to move water, and the benefits associated with water, across political boundaries.

However, the topics of the groundwater-related simulations are becoming increasingly diverse as groundwater professionals become more involved with both the technical substance of hydrogeology and the process of dispute prevention and conflict resolution. The *Groundwater Protection Dueling Expert Role Play* was developed by Jarvis (2014), a groundwater hydrologist who teaches and practices conflict resolution in groundwater and water-well construction. It provides an applied situation of the conflicts associated with multiple working hypotheses and the emerging field of scientific mediation. Likewise, the *Edwards Aquifer Case* was developed by a government scientist and academic collaborative governance practitioner for the complex situation of groundwater as a private property right grounded by the Endangered Species Act.

Board games with monies permit negotiations around a table where multiple languages are spoken. *Santiago* is a water allocation board game with farms and fleeting fidelities and that fiddles with bribery. The groundwater counterpart to *Santiago* is the *California Water Crisis Game* – a groundwater board game where the winner is the player with the best reputation.

A pioneer in computer games is the *Tragedy of the Groundwater Commons Groundwater Game* developed by IGRAC. This game is part of IGRAC's GroFutures – "Groundwater futures in Sub-Saharan Africa project." The game uses a spreadsheet model to analyze well-development impacts and economics to neighboring water users. Like the *California Water Crisis Game*, the computer-assisted *Groundwater Game* developed by the Environmental Defense Fund plays with the Sustainable Groundwater Management Act of 2014.

Save the Water Game is in development, where the first prototype has four players playing as farmers cultivating their fields, pumping groundwater from a common aquifer for irrigation to make the highest profit. The game can also incorporate roles, such as a selfish farmer, a communicative or a generous farmer, and so on. This elevates the player out of an egocentric perspective into a broader and observing perspective. The online version places the farmer against themselves.

These games directly incorporate the development history leading to the current state of aquifers, collaborative governance, the decision between individual gain and collective sustainability, or they mesh hydrogeology with law. To simulate the issues, rights, and development of aquifers over time, these components can be directly or indirectly included in the learning experience of playing an aquifer game.

Serious gaming can be used to move negotiations forward when parties have become too entrenched to form new or stronger collaborative partnerships, as discussed by Moore and Jarvis (2020). Serious gaming and scientific mediation "encourages the stakeholders to work with the information themselves and come to a consensus on what parameters they have to work within" (Moore & Jarvis, 2020, p. 49).

The *Unitization Game*, discussed in the following sections, challenges players to identify and create the new parameters to the game, even those that are not specifically identified.

11.2 The Unitization Simulation Game: Introduction

A missing component of other groundwater games has been the central role of aquifer transresources. The Unitization Game personifies the aquifer as the "mediator" of the relationships between interest groups. By giving the aquifer "personhood," the players will be able to interact with the aquifer and discover new aspects of its "hidden waters" and "hidden pore spaces."

The Unitization Game combines across-the-table negotiations, resource management, and strategy into a satisfying balancing act of human needs and environmental elements. While the wide variety of projects, technologies, and hydrology associated with aquifers has been simplified in this game, the key social and technological aspects of unitization play a central role in the simulation. Few (if any) games have the aquifer play a central role, let alone be a "player" in its own right! Let's play!

Unitization, in its most effective form, includes all stakeholders as parties to the agreement. Each party brings their own benefits to the unit, provided by consent, with the incentive as the potential value of cooperation toward a common goal. Like in real unitization, each party in this simulation contributes or has an interest in a certain aquifer transresource. Water may be the currency, but storage, water

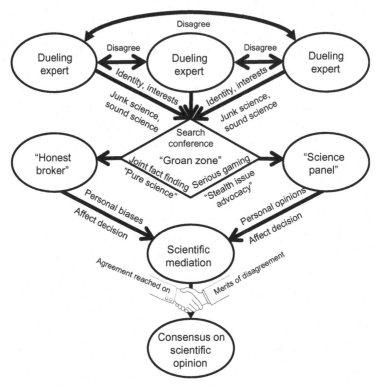

Figure 11.2 The Unitization Game

quality, temperature, and other transresources are the key to the unit's victory. As represented in Figure 11.2, players can use powers that harm, assist, or work outside the aquifer community (Shah, 2009).

There are several key unitization-related concepts included in the game, hopefully discovered as the simulation progresses:

- Storage is more important than water allocations.
- Cooperation is required to fully utilize an aquifer.
- The free-rider problem discourages participation in unitization agreements, unless unitization is compulsory.
- Operating the aquifer as a unit might not bring immediate benefits to the participant, but will improve benefits for everyone in the unit eventually.
- Cooperation and aquifer unitization can assist in overcoming the challenges of climate change, flooding, economic fluctuations, and other random events.
- Redetermination at regular intervals or after certain triggers can "reset" the game, allow new negotiations to take place, and lead to better outcomes in the long term for resource use.

11.3 Simulation Setup

1. Playing the Unitization Game should last approximately 1–1.5 h, depending on the length of discussion and negotiation.
2. Groups: Players divide themselves into six groups and select a player as the Aquifer. The six groups select which interest group they will be simulating for that year. Each interest group is given the summary of their resources, powers, and goals. *Summaries for other interest groups are not shared with all interest groups,* but interest groups are free to discuss their powers with other interest groups (if they are willing!).
3. Order: Each game consists of 5 years, when each interest group has a turn to select the power they wish to use. Due to historic development and prior appropriation, turns are in the following order: Government, Surface Water Users, Municipalities, Industrial and Commercial Users, Groundwater Users, and Environmental and Public Interest Users.
4. Initial income: The Aquifer distributes to each interest group the initial water units. This value is their water unit "income," akin to each group's allocation in a given year.
5. Initial storage: Each interest group begins with an amount of storage. The income of the interest group cannot exceed the amount of storage to which the interest group has access.

11.4 Game Rules

1. Interest group incomes cannot go below zero. For example, if the Government plays Power 1 (which decreases Surface Water Users' income by 2) and decreases Surface Water Users' income to −1, the Surface Water Users' income for the next year is recorded as 0.
2. Any interest group with a water unit income of zero cannot use a power that round.
3. At the start of their turn, each interest group selects a power. Using a power costs one water unit. If an interest group has no water units, it cannot use a power that round.
4. Agreements may be formed between any number of interest groups and can allocate water units between themselves. For example, Municipal may form an agreement with Groundwater Users to both use their Power 3 to develop new storage in exchange for allowing each to jointly use their storage. Surface Water Users may also donate unused water units to Environmental and Public Interest to use Power 2 and develop instream water rights.

5. Aquifer Storage: Aquifer storage may be utilized by any interest group that participates in an agreement that allocates the storage spaces among the interest groups. At least three interest groups must be party to the agreement to access Aquifer storage. However, any interest groups can withdraw from the Aquifer storage spaces at anytime (effectively stealing water stored in the Aquifer from other interest groups). The "stolen" water units come from the most adjacent senior interest group in priority and are limited to the number of water units for a power. If an interest group withdraws from an agreement, the water units under their "account" revert to the public domain and can be "stolen" by any interest group for their own use.

   a. For example, Groundwater Users, Municipal, and Government may agree to allocate the Aquifer storage space by splitting it three ways, but Environmental and Public Interest might withdraw stored water units from the Aquifer to use Power 2 to develop instream water rights. Environmental and Public Interest cannot withdraw more than 4 water units, since Power 4 only requires 4 water units.

6. Compulsory Unitization: During year 1 and year 2, participation in an agreement to use Aquifer storage is entirely voluntary. However, before year 3, a vote is held to determine if "compulsory unitization" should be adopted. If more than 50 percent of the interest groups vote in favor, a compulsory unitization law is passed. Under this law, if four interest groups (or more) agree to use the Aquifer, the remaining two interest groups (or fewer) are bound by the agreement (even if they don't agree to its terms). The nonparticipating interest group must be allocated at least 4 storage units in the Aquifer (determined by the participating interest groups). Of course, the two (or fewer) interest groups might determine it is in their interest to participate and negotiate for increased storage.

7. Goals: Each goal is achieved when an interest group "has control" over a certain number of water units. This does not need to be *individually* but may be part of a collective allocation under an agreement. For example, the Government and Municipal groups may have an agreement to share income and storage collectively, and at the end of year 5, their shared storage is 26 water units, meaning both achieved Goal 2.

11.5 Playing the Unitization Game in Person

1. At the beginning of each year, the Aquifer rolls a die, selecting a random event for that turn.

   a. **#1: Drought** – Each interest group is allocated 2 fewer water units that year.

   b. **#2 Litigation** – Only government and municipal interest groups may use their powers for this year.

    c. **#3 Climate change** – Permanent reduction of income of 2 water units for the rest of the game for all interest groups.

    d. **#4 Federal project investment** – Each interest group receives an additional 1 water unit for the rest of the game.

    e. **#5 Flood** – Any party receiving 3 or fewer water units reduces income by 1 (broken infrastructure). Any interest group receiving more than 3 receives an additional 1 water unit income.

    f. **#6 Economic boom/bust** – flip a coin: If heads, the interest groups receive a bonus of 2 water units of income. If tails, they receive 2 fewer water units of income.

2. The interest groups select a power (if they have the water units) and provide the cost to the Aquifer for use of that power. The powers are selected in sequence determined by priority, akin to the priority date for each groups' water right. When all six interest groups have selected and completed their turn, the year is concluded. Interest groups may discuss the game with other players during years, but cannot enter *binding* agreements with other interest groups (but nothing stops players from creating *nonbinding agreements* during a specific year).

3. Between each year, players may negotiate agreements, reallocate storage spaces in the Aquifer, or withdraw from an agreement.

4. At the end of year 5, each interest group's income and stored water units are added to determine if they have achieved one or all goals. Each interest group has a different goal but similar thresholds. If an interest group is part of an agreement that shares water incomes and/or storage, they may count all water units under the agreement toward their goal.

11.6 Playing the Unitization Game Online

Unlike many simulation games, the Unitization Simulation Game may be played entirely remotely using email, chat, or videoconferencing. While playing remotely detracts somewhat from the live negotiation and group learning aspects of the game, the turns, plays, and resource management can be conducted entirely using an email thread.

1. The Aquifer sets up the main email thread, chat, or shareable document outlining the rules and years, and summarizes how to use resources, powers, and goals. The initial email thread may also include a table somewhat like the example in Table 11.2 (at the end of this chapter) to track water units and storage. Current water units and storage volumes (marked in parenthesis) available are recorded in the table. At the beginning of the simulation and

Table 11.2. *Example game-recording table. A completed game has been entered as an example*

Game 1

Year	Random event	Government	Surface	Municipal	Industrial-commercial	Groundwater users	Environmental and public interest	Total aquifer storage
1	#1 (Drought)	Income: +4 −2 Storage: (5)	+6 −2 (8)	+3 −2 (3)	+4 −2 (2)	+4 −2 (0)	+3 −2 (2)	+10 Total group storage: (20)
Play		Power 2: +2 −2 (+0)	P1: +5 (+0)	P3: −3 (+2)	P3: +0 (+2)	P3: +0 (+5)	P1: +4 (+0)	−3 −2
2	#4 (Federal project)	+2 +1 (5)	+8 +1 (8)	+0 +1 (3)	+2 +1 (4)	+2 +1 (5)	+5 +1 (0)	+5 (25)
Play		P2: +2 (+0)	P3: +0 (+2)	P3: +0 (+2)	P3: +0 (+2)	P1: +4 (+0)	P3: −2 (+3)	−2
3	#3 (Climate change)	+5 −2 (5)	+9 −2 (10)	+1 −2 (5)	+3 −2 (6)	+5 (limit) −2 (5)	+3 −2 (3)	+3 (34)
Play		P3: +0 (+2)	P3 +0 (+2)	No units	P2: +2 +5 (+0)	P4: +0 +10 (+0)	P4: +0 (+0)	+10 +5
4	#5 (Flood)	+3 −1 (7)	+7 +1 (12)	+0 −1 (5)	+6 (limit) +1 (6)	+5 +1 (limit) (5)	+1 −1 (3)	+18 (38)
Play (unitized)		P4: +0 +10 (+0)	P4: +0 +5 (+0)	P4: +0 +10 (+0)	P4: +0 +5 (+0)	P4: +0 +10 (+0)	P4: +5 +0 (+0)	+10 +5 +10 +5

							#6 (bust)
5	+12 −2	+13 −2	+10 −2	+12 −2	+16 −2	+5 −2	+63
	(7)	(12)	(5)	(6)	(5)	(3)	(38)
Play	P4:	P4:	P4:	P4:	P4:	P4	+10
	+0 +10	+0 +5	+0 +10	+0 +5	+0 +10	+5 +0	+5
	(+0)	(+0)	(+0)	(+0)	(+0)	(+0)	+10
							+5
							+10
							+5
Totals	20	16	18	15	24	8	108
	(7) = 27	(12) = 28	(5) = 23	(6) = 21	(5) = 29(−2)	(3)	(38)
Victory?	Yes, Goal 2	Yes, Goal 2	Yes, Goal 2 (receive 2 from GW)	Yes, Goal 1	Yes, Goal 2 (provide 2 water units to surface water)	No	

before each year, the Aquifer rolls the die and selects the random event for that year (as earlier).

2. The interest groups appoint a spokesperson that communicates their final decisions to the Aquifer. Communication within interest groups can be conducted over email or video chat software.

3. In sequence, the Aquifer sends an email, chat, or asks the interest group the power they would like to use that year. Thereafter, the Aquifer sends the next interest group a similar email, chat, or asks (with the Game Recording Table updated; see Table 11.2) which power they have selected.

4. At the end of each year, the Aquifer sends the results of the year to all of the groups, and the interest groups are encouraged to negotiate with other interest groups and form agreements between years, either in a main thread or secretly using their own email thread, break-out videoconference rooms, or text messages. The Aquifer may place a time limit on the negotiation between years. After negotiation is complete, the Aquifer re-rolls for the next random event to determine the new rule that applies to that year.

5. Once the previous steps have been completed for 5 years, the game is scored in the same matter as a typical live game.

6. Participants are encouraged to play several games, learning the dynamics of the game and the goals that can be achieved, and *redetermining* their strategies for use of the Aquifer, their powers, and *allocation* of water units.

11.7 The Role of the Aquifer

The Aquifer is its own player in the Unitization Game, distinct from the Interest Groups described later. The Aquifer plays two roles in the simulation: the accountant and the objective mediator. In many ways, this play is akin to the unit operator in unitization agreements. The Aquifer suggests ideas, coordinates player agreements, manages water storage accounts, and generally enforces the agreements formed by interest groups.

The Aquifer is also its own "interest group" player, even if not recognized as such. In this sense, the Aquifer is given environmental personhood and granted some of the same right to score as the human players in the game. The Aquifer is free to advocate for itself throughout the game and receives its own score. Unfortunately, the Aquifer is a "passive" player in the Unitization Game, as the Aquifer's destiny is in the hands of the interest groups and subject to the human players' ability to cooperate.

11.8 Summaries of Interest Groups

11.8.1 Government

Resources:
- Initial income: 4 water units
- Initial storage: 5 water units

Powers:
- Power 1: Add 3 to Government income, decrease Surface Irrigator income and Aquifer storage by 2, by requiring conversion from flood to sprinkler or drip irrigation.
- Power 2: Add 2 to Government income by funding atmospheric water collection technologies.
- Power 3: Add 2 to Government storage by raising dam heights and installing fish ladders.
- Power 4: Add 5 to Environmental and Public Interest income and add 10 units to Aquifer storage by studying the Aquifer by awarding grants to install subsurface water-cooling systems for power plants and cooling discharges to surface streams.

Goals:
- Goal 1: Have access to 10 water units to offset groundwater uses contaminated from leakages from military bases.
- Goal 2: Have access to 25 water units to improve scenic areas and increase tourism.
- Goal 3: Have access to 35 water units as a reserve supply in case of national emergencies and climate catastrophes.

11.8.2 Surface Water Users

Resources:
- Initial income: 6 water units
- Initial storage: 8 water units

Powers:
- Power 1: Add 5 units to Surface Water User income and decrease Municipal and Aquifer storage by 3, by replacing earthen ditches with pipelines and preventing incidental aquifer recharge.
- Power 2: Add 2 to Surface Water Users' Income, by importing water from an adjacent basin.

- Power 3: Add 2 to Surface Water Users' Storage, by constructing small-scale storage ponds.
- Power 4: Add 10 to Government income and add 5 units to Aquifer storage, by studying the Aquifer; recapture contaminated irrigation water for beneficial use and reduce groundwater remediation needs.

Goals:
- Goal 1: Have access to 10 water units by year 5 to mitigate conveyance losses seepage and increasing crop needs due to increasing temperatures, decreased precipitation, and climate change.
- Goal 2: Have access to 25 water units by year 5 to construct a new reservoir to store winter flows.
- Goal 3: Have access to 35 water units by year 5 to raise the groundwater table to decrease seepage and restore source water streamflow.

11.8.3 Municipal

Resources:
- Initial income: 3 water units
- Initial storage: 3 water units

Powers:
- Power 1: Add 3 to municipal income and decrease Industrial-Commercial income and Aquifer storage by 2, by requiring retrofits for water efficiency and decreasing unintended leaks and discharges.
- Power 2: Add 2 to Municipal income by developing new water sources
- Power 3: Add 2 to Municipal storage by developing new surface reservoirs.
- Power 4: Add 5 to Surface Water Users' income and add 10 to Aquifer storage, by studying the Aquifer and constructing a subsurface dam within city limits.

Goals:
- Goal 1: Have access to 10 water units to mitigate current overdevelopment due to lack of concurrency rules.
- Goal 2: Have access to 25 water units for anticipated growth in the near future.
- Goal 3: Have access to 35 water units to prevent subsidence and damage to infrastructure by injecting water strategically to prevent permanent aquifer compaction.

11.8.4 Industrial and Commercial

Resources:

- Initial income: 4 water units
- Initial storage: 2 water units

Powers:

- Power 1: Add 3 to Industrial and Commercial income and decrease Groundwater Users' and Aquifer storage by 2, by increasing wastewater discharges, by developing high-capacity wells and causing subsidence (aquifer inelastic aquifer system compression).
- Power 2: Add 2 to Industrial and Commercial income by purchasing imported water.
- Power 3: Add 2 to Industrial and Commercial storage by constructing a tank farm.
- Power 4: Add 10 to Municipal income and add 5 to Aquifer storage by implementing an in situ bioremediation prefilter for effluent.

Goals:

- Goal 1: Have access to 10 water units by year 5 to offset uses harmed by groundwater contamination.
- Goal 2: Have access to 25 water units to inject water and control groundwater pollutant movement.
- Goal 3: Have access to 35 water units to pump out, treat, and replace contaminated groundwater.

11.8.5 Groundwater Users

Resources:

- Initial income: 4 water units
- Initial storage: 0 water units

Powers:

- Power 1: Increase Groundwater Users' income by 4 and decrease Environmental and Public Interest income and Aquifer storage by 2, by litigating conjunctive management rules and defending high-rate groundwater pumping (and subsidence).
- Power 2: Add 2 to Groundwater Users' income by drilling new wells.
- Power 3: Add 5 to Groundwater Users' storage by installing small-scale artificial storage and recover projects.

- Power 4: Add 5 to Industrial-Commercial income and add 10 to Aquifer storage by providing data for computer modeling and participating in centralized wellfield operations.

Goals:
- Goal 1: Have access to 10 water units by year 5 to offset the sinking water table.
- Goal 2: Have access to 25 water units for increasing irrigation rates and develop new subdivisions.
- Goal 3: Have access to 35 water units to restore the water table for domestic and irrigation wells, restore natural discharges to surface waters, and maintain a reserve supply in case of drought.

11.8.6 Environmental and Public Interest

Resources:
- Initial income: 3 water units
- Initial storage: 2 water units

Powers:
- Power 1: Add 4 to Environmental and Public Interest income and decrease Government income and Aquifer storage by 2, by locating a keystone species and establishing access to water and prohibiting well development under an Endangered Species Act's biological opinion.
- Power 2: Add 3 to Environmental and Public Interest income by filing for instream water rights.
- Power 3: Add 3 to Environmental and Public Interest storage by constructing artificial wetlands.
- Power 4: Add 10 to Groundwater Users' income and add 5 to Aquifer storage by strategically placing artificial wetlands in recharge zones with high infiltration rates.

Goals:
- Goal 1: Have access to 10 water units by year 5 to offset the sinking water tables and provide water to shrinking groundwater-fed wetlands.
- Goal 2: Have access to 25 water units for augmenting stream flows in summer and fall for fish migrations.
- Goal 3: Have access to 35 water units for restoring the water table, natural source springs for wetlands, and protect rare subterranean fauna and microorganisms.

11.9 Scoring the Simulation Using Rationalization Scenarios

As opposed to the complexity of the Unitization Game, the scoring system is quite simple. Human players and the Aquifer each receive a score at the end of each year.

For the Aquifer, there can be three outcomes using the rationalization scenarios (Figure 11.3) by Foster Nanni et al., (2003). These three scenarios are reflected in the four powers held by each interest group, which determine the Aquifer's destiny. Since each interest group is free to choose when it uses its powers (at least in the beginning without a multilateral unitization agreement), each interest group will likely have had its own interests in mind at the expense of others (and the Aquifer!)

To score the Aquifer, the average of the powers is determined and rounded to one of the three outcomes below depending on the score range. To determine the score, add up the total value of the powers used by players (Power 1 = 1, Power 2 = 2, Power 3 = 3, Power 4 = 4) during the game and divide by 30 (assuming 5 years per game). As in the real world, the Unitization Game does not emphasize a certain outcome. Instead players choose the Aquifer's destiny through their selection of powers that directly or indirectly affect the aquifer.

- **Score ranges 1–1.9 (Gradual Depletion):** Interest groups fully invest in production and extraction without regard to impacts on neighboring interest groups and the Aquifer. Returns on investment primarily benefit the interest group at the cost of other interest groups, but those benefits are limited and cannot achieve all

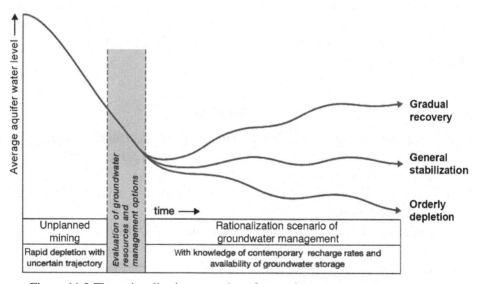

Figure 11.3 The rationalization scenarios of groundwater management

goals. The Aquifer has compacted irrecoverably and will become less useful as time progresses, as water quality diminishes. Some interest groups may have achieved their goals, but at the cost of economic and social damage to other interest groups.

- **Score ranges 2.0–2.9 (General Stabilization):** Interest groups have achieved stability by mitigating damage and balancing their water demands with other sources. However, interest groups have not restored the benefits the Aquifer once provided, and have only slowed or stopped damage. Some interest groups have achieved their goals, while others have not.
- **Score ranges 3.0–4.0 (Gradual Recovery):** Interest groups leverage the indirect benefits of working with other interest groups and coordinate the use of the Aquifer, benefiting all players. The Aquifer has been developed and studied, and interest groups have cooperated to jointly manage the system.

For the human players, the score is determined by adding the total value of the goals achieved (Goal 1 = 1, Goal 2 = 2, Goal 3 = 3) and dividing by 18. This percentage is akin to the overall social benefits achieved during that game. For example, a score of 50 percent means that only 50 percent of the potential social benefits of the game were achieved.

11.10 Discussion Questions

At the end of each game, the players answer the following discussion questions. These questions invite the players to reflect on the simulation and how the interest groups leveraged their powers. This also allows the players to develop new strategies and explore the dynamics that come with developing an aquifer unitization agreement.

11.10.1 After Game 1

1. What was the main reason that interest groups, if any, were able to achieve their goals? Was the outcome the best social outcome? Or did players mostly concern themselves with their own goals?
2. How did the random events affect the game? Did they severely hinder any specific interest groups?
3. Did some interest groups have any unfair advantages? Did any groups have any unfair disadvantages? How did these advantages/disadvantages influence their choices in the game?
4. Did interest groups form any agreements/deals to share storage spaces or water units? Did they also agree to only use specific powers?

5. Did the Aquifer play a large role in the game? Did players consider the impact on the Aquifer and the resulting rationalization scenario during gameplay?

11.10.2 After Game 2

1. Compare the last game with this game. What changed? What didn't? Did more players achieve their goals? If so, why?
2. Did risk play any role in the game, whether in negotiation of agreements or in the selection of which power to use? Did the players discover that assisting players earlier in priority (using Power 4) was riskier or less risky than harming players later in priority (using Power 1)?
3. Did the random events have more or less of an effect in this game? Did the agreements, if any, assist with overcoming these random events?
4. What was the biggest challenge to forming agreements? Were there any parties that refused to participate and took advantage of the access to the Aquifer?
5. Does the group of players want to pass a "compulsory unitization" law that requires all parties to be part of any unitization agreement, even if they don't agree?

11.10.3 After Game 3

1. Was every interest group able to achieve their Goal 3? How did the players accomplish this feat?
2. Were the players better able to withstand the effects of random events this game? Did the agreement, if any, provide any assistance?
3. What parts of this simulation do you think represent reality? And which parts do you think could be improved? Do you think there is a missing aspect of aquifer unitization not represented in the game?
4. Compared to the first and second game, what strategies did the interest groups deploy to achieve their goals? Did the interest groups include other interest groups' goals into their strategy?
5. What was the biggest challenge to cooperation? Was the initial allocation of resources the primary factor? Or did personality and relationships play a bigger role?

11.11 Example Game

The following example game shows how the Unitization Game could simulate changes in water unit income and storage. In the example game, the interest groups began using Powers 1–3, attempting to increase their water unit incomes and storage, neglecting the impacts on the Aquifer.

In the "Year" row, the random event is recorded and the impact it had on the water income and/or storage is noted under each interest group. In the "Play" row, the power used by each interest group is noted along with the corresponding changes associated with the power selected by the interest group. The "Total Aquifer Storage" column tracks the Aquifer's storage and the total storage for all the players combined.

In this example game, the interest groups negotiated a unitization agreement in Year 4, whereby each interest group agreed to play their Power 4 in exchange for access to the Aquifer's storage, sharing any water units at the end of the game, and agreeing to maximize the goals achieved by redistributing water units at the end of the game (given to the earliest interest group in priority). In this case, the Groundwater User's extra water units were provided to Municipal rather than Industrial-Commercial due to their priority. The Environmental and Public Interest Group almost achieved its Goal 1, if only the economy had not busted!

Table 11.3. *Example Aquifer scoring*

Year	Powers total
1	2 + 1 + 3 + 3 + 3 + 1 = 12
2	2 + 3 + 3 + 3 + 1 + 3 =15
3	3 + 3 + 0 + 2 + 4 + 4 =16
4	4 + 4 + 4 + 4 + 4 + 4 = 24
5	4 + 4 + 4 + 4 + 4 + 4 =24
Score	91/30 = **3.03**

Table 11.4. *Interest group scoring*

Interest group	Goal achieved?
Government	Goal 2
Surface water users	Goal 2
Municipal	Goal 2
Industrial-commercial	Goal 1
Groundwater users	Goal 2
Environmental and public interest	None
Score	9/18 = **50%**

To score this game for the Aquifer, the sums of the values of each power are combined and divided by 30 to provide the outcome for the Aquifer. The calculation is represented in Table 11.3. Since the Aquifer is often an afterthought until the interest groups realize its value, the Aquifer had the hardest time "wining" the game. In this case, the Aquifer is *barely* able to achieve a gradual recovery.

To score the game for the interest groups, the sum of the goal value is determined and divided by 18 to determine the social benefits achieved by the end of the game. This calculation is represented in Table 11.4. In this case, only 50 percent of the potential social goals were achieved.

12

Conclusions and Recommendations for Future Research

> Unitization certainly did work in the oil and gas context. While it was
> fought by some, it has proven to be the savior of all.
>
> —*S. E. Clyde (2011)*

The myths of aquifer science have been mostly dispelled, but the mysticism of governance still lurks behind current approaches to sustainable groundwater use. Aquifers are combinations of both public and private rights, physically bound, and must be treated as elements in a complete aquifer system. The examples described in this book show that progress has been made to move from groundwater management to groundwater governance, but in only some cases has governance reached the aquifer. Unitization of oil and gas fields provides a valuable analogy to aquifers, showing an approach that bridges the science with the social and legal context, forming a negotiated, often voluntary governance system. As shown in this book, the unitization contract system could potentially be applied to aquifers across the United States and abroad.

While this book described the application and use of CAGAs within a single state and legal system, unitization in oil and gas has been used internationally and in transboundary contexts. Future research into these examples could provide a model for the approach's use across state and national boundaries with different systems of rights and legal requirements.

However, many of the aspects of these international versions of unitization remain the same, since these legal issues are subcomponents in the eventual determination of shares and negotiated goals of the agreements. Unitization has been successful, becoming the accepted norm in oil and gas reservoir governance.

Unitization's success has also led to its biggest challenges. As a collective property approach, it is often subject to legal attack as a form of monopoly. As discussed by Wiley (2019), Weaver (1986), and Scott (2016), these kinds of collective governance systems can sometimes be subject to antitrust legislation as anticompetitive forms of association. As Scott (2016) suggests, courts are unlikely to find an organization anticompetitive when the purpose is the improvement of social and shared economic value. Further, unitization has proven too beneficial and survived despite antitrust litigation (Weaver, 1986). Further research into collective governance organizations and their interaction with antitrust law could provide additional support for the use of unitization despite the threats of anticompetitive behavior.

Wiley (2019) discusses that social concerns might also interfere with the successful negotiation of an aquifer unitization agreement, like the issues of free riders undermining the effective operation of a unit or perceptions of a lack of public interest and environmental review. The issue of free riders could be resolved with the coordinated use of irrigation district laws or the adoption of compulsory unitization statutes. Jarvis (2014) identifies that public interest groups could be directly incorporated into a unitization agreement, potentially alleviating concerns about public interest oversight. Further, unitization would remain within the jurisdiction of regulatory agencies that currently provide public interest oversight. Still, additional research could investigate how a proposed aquifer unitization project could be perceived by the public, regulators, and legislators.

Another area of potential research uses a similar governance approach to unitization: the corporation. Corporations use many of the same components as unitization (shares, classes of shares, shareholders, boards of directors, bylaws, and private rule systems). The corporate model could use many of the same suggestions for shares, classes of shares, and elections as provided in the CAGA governance approach. Instead of shares being equitable interests in a contract, shares should be ownership interests in an aquifer corporation, potentially with classes of shares representing components of storage and the aquifer mass balance equation. Dividends would be water allocations rather than corporate profits. Corporations could be a more effective legal form of aquifer governance, allowing further investment than would typically be available in a CAGA (being limited to present aquifer users). Alternative forms of corporations (B-Corporations) merge profit-seeking motives with benefits to communities, the economy, and the environment.

These variations on governance systems could also form the foundation for a CAGA-like legal structure. Whether water is private or public property, innovative and flexible solutions to aquifer governance are desperately needed. Either through private contract, public policy, or legal argument, groundwater governance should transition into aquifer governance. The current "policy drought" has left our aquifers ineffectively governed and unscientifically managed (Anderson, 1983). New solutions that bridge the legal and social aspects and the science of aquifers can create a socially and environmentally sustainable future for aquifers (Faigman, 1999).

12.1 Unitization and Integrated Water Resources Management?

Looking forward, does unitization and collective action of groundwater, aquifers, and the interrelated resources, coupled with the multiple uses of the subsurface, fit into the process of integrated water resources management (IWRM)? Foster and Ait-Kadi (2012, p. 417) asked the important question as to where groundwater fit in IWRM by first reminding the reader that IWRM "is the process of managing water resources holistically and of promoting coordinated consideration of water, land and related natural resources during developmental activity." And to implement groundwater into the IWRM process, they suggest that the approach needs to be "multidisciplinary, strongly participatory and bridge across sectors." Through their work with the Global Water Partnership, Foster and Ait-Kadi (2012) indicate that groundwater management will require an integrated approach to (1) the land–water management interface in the interest of conserving groundwater recharge and quality; and (2) spatial allocation of resources to different uses (including ecosystems) than are usually attempted in river-basin management. For really extensive and deep aquifers, they suggest a decentralized approach to IWRM is preferred for these groundwater systems.

In discussing the nexus between the Watercourses Convention and the Law of Transboundary Aquifers that may eventually become part and parcel of implementing IWRM, Carlson (2011) recommended that states be permitted the flexibility to self-organize through bilateral and multilateral agreements, typical of the principles of unitization of transboundary oil and gas reserves, and that "any affirmative obligations should be based on the need to protect the human right to water," which ought to be explicitly stated in the Law of Transboundary Aquifers (Carlson, 2011, p. 1436). The application of unitization principles to groundwater and aquifers may be the "pragmatic framework for integrated action" that Foster and Ait-Kadi (2012, p. 417) were searching for, including linking to the human rights–based approach for integrating the subsurface into IWRM.

We feel that one of the missing pragmatic frameworks for integrated action could be CAGA. A survey of best practices of thirty-one unitization provisions in oil and gas states within the United States has important advice for application to groundwater and aquifers: "Fieldwide unitization, even with statutory assistance and for all its logic, belongs to that ... special class of legal endeavors for which success can result only when all participants *practice good faith, know their subjects, and govern their actions with wisdom*" (Eckman, 1973, p. 381, emphasis added).

Appendix

Model Collective Aquifer Governance Agreement

Disclaimer: *This conceptual model contract has been prepared only as a theoretical guide and may not contain all the provisions that may be required by the parties to an actual agreement. Jakob Wiley is not an attorney and does not endorse its use as a contract. Use of this model contract or any portion or variation thereof shall be at the sole discretion and risk of the user parties. Users of the model contract or any variation thereof are encouraged to seek the advice of legal counsel to ensure that the final document reflects the actual agreement of the parties. Jakob Wiley disclaims any and all responsibilities or liability whatsoever for loss or damages that may result from use of this model contract or portions or variations thereof. The major components and organization of this model agreement are modified version of the Model Unitization and Unit Operating Agreement (2006) developed by the Association of International Petroleum Negotiators and used with permission.*

Contents

1 Purpose

WHEREAS, the purpose of this agreement is the establishment of a continuing aquifer governance system;

WHEREAS, the participating landowners, organizations, and aquifer users wish to organize efforts for aquifer governance and sustainable development;

WHEREAS, participating landowners, organizations, and aquifer users desire to utilize the expertise for technical, organizational, and legal governance of the aquifer; and

THEREFORE, these individuals and groups wish to form a collective aquifer governance agreement.

2 Definitions

A. Agreement means the Collective Aquifer Governance Agreement.

B. Unit Operator means the person, organization, corporation, law firm, or other entity selected to manage the Shareholders' aquifer.

C. Aquifer Basin means the region containing the aquifer.

D. Aquifer means the portion of the subsurface that contained, contains, or could contain, water within pore spaces.

E. Unit Boundary ("UB") means the region composed of participating Landowners' properties.

F. Shareholder means any person, landowner, organization, or other entity that holds shares in the Agreement.

G. Non-shareholder means any person, landowner, organization, or other entity that is not a member of this agreement.

H. Shares are equitable interests in the Agreement as described later.

I. Unit Plan ("Plan") means the document that outlines the water management priorities, improvement projects, and financial information for the Unit Operator's management of the UB.

J. Project means any aquifer recharge, injection, recovery, improvement, or rehabilitation operation conducted by the Unit Operator.

K. Exploration Activity means any exploratory well, geologic testing, or any other research on the aquifer or subsurface located in the UB.

L. Person means an individual, corporation, company, partnership, limited partnership, limited liability company, trust, estate, Government agency, or any other entity.

M. Share Value means the monetary value of a single Share.

N. Expenditure means the cost to construct, maintain, and operate any Project.

O. Decommissioning means the dissolution of the Agreement and the cease of all governance activities by any Unit Operator.

P. Non-Unit Activity means any activity occurring on property within the UB that does not relate to groundwater, pore spaces, or aquifer governance.

Q. Decommissioning Date is the date of the vote to terminate this Agreement.

R. Effective Date means the _____ or the date when all parties have signed the agreement, whichever occurs later.

3 The Committee

A. The Committee shall be composed of the Shareholders listed in Exhibit _____ (number).
B. The Committee shall be composed of at least fifteen (15) Shareholders who volunteer to participate in the selection of the Unit Operator.
C. Volunteers for the Committee shall contact other Shareholders who wish to be members of the Committee within thirty (30) days of the Effective Date.
D. New members of the Committee may be added by vote of greater than 50 percent the Committee.
E. Each Committee member shall select an alternative party or agent to take their place in the case of an absence.

4 Unit Operator Selection Procedure

A. One Share represents one vote in the election of the Unit Operator.
B. The Committee shall publish and advertise for potential MOs until _____ (date). A list of potential MOs shall be collected. Organizations that contact the Committee after _____ (date) shall not be considered as a potential Unit Operator.
C. Selection Process.
   i. The selection of the Unit Operator occurs on _____ (date) or nine (9) years and six (6) months after the Effective Date.
   ii. On the selection date, the Committee shall tally the votes publicly and post the number of votes for each potential Unit Operator.
   iii. The Committee shall mail a letter three (3) months before the selection date informing each Shareholder of the selection process, all potential Unit Operators available for selection, the date of the selection, the location of the selection, the number of votes for the letter's recipient, the total number of votes in the UB, and the address to which Shareholders shall mail their voting letters.
   iv. The Committee shall collect and store voting letters securely until the selection date.
   v. Voting letters shall inform the Committee of the Shareholders' desired Unit Operator.
   vi. Shareholders shall mail their voting letters to the address provided in the letter described in Section 3(F)(ii).
   vii. Voting letters received by the Committee shall not count in the selection of the Unit Operator.

Acceptance.
   i. The Committee shall publish an Election Report containing the selection results and mail a copy to each Shareholder and each potential Unit Operator.
   ii. The Committee shall provide the Agreement to the potential Unit Operator with the highest vote tally.

iii. If the selected Unit Operator refuses to sign the agreement, the Committee shall provide the Agreement to the next-highest vote recipient.

iv. If the next-highest vote recipient refuses to sign the Agreement, this Agreement is dissolved.

v. The Committee shall begin preparation for the next selection nine (9) years after the Unit Operator signs this Agreement.

5 Removal and Resignation of Unit Operator

A. The Unit Operator may be removed by a petition of Shareholders.

B. If the Committee receives a petition to remove the Unit Operator with greater than 66 percent of all shareholders signing, the Committee shall organize a new selection as described in Section 4 within sixty (60) days of receiving the petition.

C. The removed Unit Operator shall cease all operations in the UB and shall hold all Shareholders, the Committee harmless for any damages liability or losses caused by removal from this Agreement. This Section does not pertain to obligations under other agreements between Shareholders or between any Shareholder and the Unit Operator.

D. All groundwater models and data shall be provided to the Committee in a usable and accessible form within sixty (60) days of receiving the petition.

E. Resignation of the Unit Operator.

i. The Unit Operator may resign its position with _____ (time) notice to all Shareholders.

ii. The Committee shall meet as soon as practicable to appoint a successor Unit Operator pursuant to the voting procedure above. No Party may be appointed successor Unit Operator against its will.

6 Duties and Responsibility of the Unit Operator

A. The Unit Operator shall operate the UB for ten (10) years after the Unit Operator signs the agreement.

B. The Unit Operator shall develop hydrologic models of the aquifer within the UB, including a water budget of surface and groundwater supplies, and storage potential in pore spaces.

C. The Unit Operator shall research, develop, and organize funding for potential groundwater development and improvement projects in the UB.

D. The Unit Operator shall make all decisions regarding the selection of design, data interpretation, and water management decisions for projects.

E. The Unit Operator must obtain consent for Projects that encumber Shareholders' property rights.

i. MOs may lease land or pore spaces from Shareholders or other nonparticipating organizations. These leases shall be separate agreements and are not a component of this Agreement. Parties to any lease shall be the Shareholders and the individual Shareholder owning the encumbered property. The Unit Operator may represent the Shareholders in any lease.

ii. MOs shall not obtain easements or other permanent property interests in Shareholders' real property for the development or operation of a Project.

F. The Unit Operator shall mail a copy of the current Plan to all shareholders annually.

Operating Plans

    i. The Unit Operator shall create a Plan within one (1) year after the Unit Operator signs this agreement.

    ii. The Plan shall be mailed to each Shareholder earlier than one (1) year and fourteen (14) days after Unit Operator signs this agreement.

    iii. The Plan shall contain the Projects proposed by the Unit Operator, a review of the design and location of the Projects, budgets and financial information for the Projects.

    iv. The Plan shall contain the water budget for the UB and projections for the next ten years under current and future projects.

    v. The Plan shall include total water volumes in the UB and total pore spaces and storage potential in the UB.

    vi. The Plan shall provide information regarding the negotiations with non-Shareholders and any litigation regarding the Unit Operator's use of the UB.

    vii. The Plan must include all changes and amendments from previous versions of the Plan.

    viii. The Plan shall provide a map of the UB, all Projects, and an updated three-dimensional graphic representation of the aquifer.

    ix. The Plan shall include a section describing any aquifers, water sources, or aquifer data discovered in exploration activities in the previous year.
        The Plan shall include information about any relocation of Groundwater Shares or Storage Shares.

    x. The Plan shall include water quality information for the UB.

G. The Unit Operator shall represent any Shareholder in any litigation or administrative actions before a state or federal agency.

H. The Unit Operator, in the name of a shareholder, may apply for a water right, temporary transfer of a water right, injection permit, or abandonment of a water right with written consent of the Shareholder.

I. The Unit Operator may apply for grants, loans, or other funding mechanisms for Projects.

J. The Unit Operator may propose a water charge, per capita fee, or other fee structure for groundwater use in the UB.

    i. A water charge, per capita fee, or other fee structure proposal must be mailed to each Shareholder.

    ii. The payment schedule, the rate, and expected income from the water charge, per capita fee, or other fee structure shall be provided in the proposal.

    iii. The Unit Operator must receive letters agreeing to the charge from greater than 66 percent of the Shareholders within sixty (60) days before requiring any kind of charge described in Section 4(L).

    iv. The Committee shall tally the letters proposing a charge or fee after sixty (60) days but before ninety (90) days after the proposal for a charge or fee and inform the Unit Operator and Shareholders of the outcome by mail.

K. The Unit Operator may require annual water use reports from Shareholders.

L. The Unit Operator may contract with other organizations for the construction, maintenance, and operation of Projects.

M. The Unit Operator may contract with other organizations for the fulfillment of obligations and duties under this Agreement.

N. The Unit Operator may use geologic exploration techniques to locate aquifers, water sources, and geologic information within the UB.

O. The Unit Operator shall operate and maintain all Projects and facilities located in the UB.
P. The Unit Operator shall bear liability for any activities of parties contracted by the Unit Operator on Shareholder property within the UB.
Q. The Unit Operator shall have adequate insurance for all activates conducted by the Unit Operator in the UB.

7 Regional Groundwater Contracting

A. The Unit Operator may contract with non-participating people, organizations, or municipal organizations for water conveyance, storage, supply, or other water-management project.
B. Income generated by the Unit Operator shall be divided among Shareholders using the following methods:
    i.   Income from the removal of water from the UB shall be divided among Shareholders in proportion to the volume of water removed from pore spaces within the landowner's property, after subtracting operating costs.
    ii. Income from the storage of water in the aquifer shall be divided among the Shareholders in proportion to the volume of pore space occupied by stored water, after subtracting operating costs.
    iii. The volumes of water removed or occupied pore spaces shall be based on the aquifer model developed by the Unit Operator described in Section 4(B).
    iv. The Unit Operator shall provide an itemized invoice for all payments made to Shareholders with the payment. Payments shall be mailed to the Shareholder within thirty (30) days of the Unit Operator receiving the income.
C. Any cost for Projects and other contracts shall come from funding sources described in Section 4(J) or Section 4(K) or operating costs described in Section 5(B).

8 Shareholder Obligations

A. Shareholders shall allow access to Projects located on their property. The Unit Operator shall inform the Shareholder by telephone, text message, letter, or electronic communication when access is needed for greater than one (1) hour.
B. Shareholders shall pay any charge or fee under Section 5(B) within fourteen (14) days after it is mailed. The Unit Operator shall send a late payment notice after twenty (21) days without payment.
C. Shareholders shall provide water-use data as described in Section 4(L) by the date determined by the Unit Operator.

9 The Unit Boundary

A. The Unit Operator shall operate, maintain, and manage the Projects within the UB.
B. The Unit Operator shall determine the quantity of stored water, pore space, and usable portion of the aquifer for each property included in the UB.
C. The UB extends to the maximum depth into the subsurface recognized by law as the property of the overlying landowner.
D. The UB includes groundwater resources, pore spaces, and other subsurface features of the subsurface located within the Shareholder's property.

E. Any increase in the UB by purchase of additional property requires the written approval of all Parties to the Agreement.
F. Any modification, addition, or expansion of the UB shall automatically reallocate Share distributions to Shareholders in the UB.
G. This agreement is intended to run with the land and remain binding on future land-owners in the UB.

10 Shares

Option #1: Land Area

A. All Shares as of the Effective Date shall be as set forth in Exhibit
_____ (number).
B. Shares held by each Shareholder shall be rounded to the nearest
_____ decimal place, with digits of five and above rounded upward and those of four and below rounded downward.
C. Shares are allocated based on the number of acres of land owned by a participant. One share is the equivalent of every one-hundred (100) square feet of surface property.
D. Shares remain associated with the property and do not transfer if the property is sold.
E. The Unit Operator shall use computerized surveying techniques to determine each Share.

Shares are reallocated annually by the Unit Operator.

Option #2: Pore Space Volume Basis

A. All Shares as of the Effective Date shall be as set forth in Exhibit
_____ (number).
B. Shares held by each Shareholder shall be rounded to the nearest
_____ decimal place, with digits of five and above rounded upward and those of four and below rounded downward.
C. Groundwater Shares are allocated based on the volume of water present in pore spaces below a Participant's property.
D. Storage Shares are allocated based on the volume of unfilled pore spaces below a Participant's property.
E. Groundwater Shares and Storage Shares are determined at the time of the signing of this Agreement.
F. If the Unit Operator discovers a new source of groundwater or pore space in the UB, the Unit Operator shall reallocate Groundwater Shares and Storage Shares.
   i. The reallocated shares represent the volumes of water and pore space at the time of the signing of this Agreement accounting for the newly acquired information.
   ii. The Unit Operator shall inform those Shareholders that could have reallocated Groundwater Shares and Storage Shares within thirty (30) days by mail after the discovery.
   iii. When the Unit Operator has reallocated Groundwater Shares and Storage Shares, the Unit Operator shall inform all Shareholders of the changes and the new total number of Shares in the UB.
   iv. The Unit Operator shall include Redetermination information in the annual Plan.

G. Groundwater Shares and Storage Shares remain associated with the property and do not transfer if the property is sold.

Option #3: Volumetric Water Right Conversion Basis

A. All Shares as of the Effective Date shall be as set forth in Exhibit _____ (number).
B. Shares held by each Shareholder shall be rounded to the nearest _____ decimal place, with digits of five and above rounded upward and those of four and below rounded downward.
C. Shares are allocated by the annual withdrawal volume allowed by water rights and domestic water uses.
D. One share is equivalent to one acre-foot per year as described in a water right application, permit, or determined by court decree, order, or another legal source.
E. Groundwater Shares and Storage Shares remain associated with the property and do not transfer if the property is sold.

Option #4: Mass-Balance Basis

A. All Shares as of the Effective Date shall be as set forth in Exhibit _____ (number).
B. Shares shall be determined by each water rights' contribution to the mass balance of the aquifer system.
C. Shares shall be distributed in a series of classes.
D. Groundwater rights shall be converted into shares of storage depletion, induced recharge (stream capture), evapotranspiration, and artificial recharge.
E. Surface water rights shall be converted into shares of artificial recharge, evapotranspiration, and return flows.
F. The Unit Operator shall determine the conversion rate between share classes based on the best available information.

Option #5: Water Market Basis

A. All Shares as of the Effective Date shall be as set forth in Exhibit _____ (number).
B. Shares shall be distributed in two classes: present storage and annual allocations.
C. Present storage shares shall be determined by estimation of the current groundwater stored within a Shareholder's land
D. Annual allocations shall be determined by the Unit Operator based on hydrologic conditions and the goals of the Unit Plan.
E. Shares are reallocated every _____ (time interval).

Option #6: Quality, Geothermal, or Biological Basis

A. All Shares as of the Effective Date shall be as set forth in Exhibit _____ (number).

B. Shares held by each Shareholder shall be rounded to the nearest _____ decimal place, with digits of five and above rounded upward and those of four and below rounded downward.

C. Shares shall be determined by the aggregate (concentration of contaminants, localized level of toxicity, geothermal heating capacity, biological processing rate, etc.) within a Shareholder's property.

D. The Unit Operator shall determine the (concentration of contaminants, localized level of toxicity, geothermal heating capacity, biological processing rate, etc.) using computerized groundwater models.

E. Shares are reallocated every _____ (time interval).

11 Payments and Expenditures of Shareholders

A. The Unit Operator shall make payments to Shareholders for net income generated by the Unit Operator's activities in the UB. Net income is the total income minus the operating costs of the Unit Operator.

B. The Unit Operator shall make payments to Shareholders on the _____ (date) annually.

C. Share Values shall be determined by dividing the total net income of the Unit Operator generated in the UB by the total number of Shares in the UB. The Unit Operator shall pay each Shareholder the Share Value of the Shares possessed by the Shareholder.

D. All Expenditures shall be divided proportionally among the Shareholders by the number to Shares possessed by each Shareholder.

12 Withdrawal

A. Shareholders not in default shall have a right to withdraw partially or completely from this agreement by notifying the Unit Operator and all other Shareholders in the UB.

B. Any withdrawing Shareholder shall maintain their portion of the costs and payments due to their Share until the date of withdrawal.

C. Any withdrawing Shareholder retains liability and responsibility for the costs of Project removal or other expenses for Unit Operator equipment removal from the Shareholder's property for which the Shareholder was required to contribute.

13 Aquifer Governance Guidelines and Goals

Option #1: Managed Recharge

A. The Unit Operator shall manage the aquifer to increase the volume of water stored in the UB. The Unit Operator shall research and develop Projects to increase the total volume/head/pressure of water in the aquifer to _____ (volume/head/pressure) by the year _____ (date).

B. The Unit Operator shall determine in the Plan the Projects (if any), water charge rate (if any), or other methods to increase the stored volume of groundwater in the UB.

Option #2: Sustainable Use

A. The Unit Operator shall manage the aquifer to maintain the volume of water stored in the UB. The Unit Operator shall research and develop Projects that balance the annual average groundwater extraction with groundwater recharge.

B. The Unit Operator shall determine in the Plan the Projects (if any), water charges (if any), or other method to maintain the current groundwater conditions in the UB.

Option #3: Managed Depletion

A. The Unit Operator shall manage the aquifer to reduce the volume of water in the UB. The Unit Operator shall research and develop Projects that will maximize the extractable amount of groundwater in the UB in the next _____ years and fully dewater the UB.

   The Unit Operator shall determine in the Plan the Projects (if any), water charges (if any), or other method to efficiently extract all groundwater practicable.

Option #4: Aquifer Market and Banking

A. The Unit Operator shall manage the aquifer to maximize the Share Value or Number of Share Allocations of the Shareholders in the UB. The Unit Operator shall research and develop Projects that will improve the utilization of the UB for groundwater storage and extraction.

B. The Unit Operator shall determine in the Plan the Projects (if any), water charges (if any), or other methods to expand and improve the storage utility of the UB.

Option #5: Subsidence Abatement

A. The Unit Operator shall manage the aquifer to reduce the levels of subsidence in the UB and maintain the subsurface effective stresses of the UB. The Unit Operator shall research and develop Projects that will prevent surface elevation decreases due to groundwater extraction.

B. The Unit Operator shall determine in the Plan the Projects (if any), water charges (if any), or other methods to prevent subsidence in the UB.

Option #6: Prevent Saltwater Intrusion

A. The Unit Operator shall manage the aquifer to prevent saltwater intrusion into the UB. The Unit Operator shall research and develop Projects to hinder or prevent saltwater migration into the UB.

   The Unit Operator shall determine in the Plan the Projects (if any), water charges (if any), or other methods to prevent saltwater contamination of the UB.

Option #7: Water Quality Improvement

A. The Unit Operator shall manage the aquifer to reduce the harmful results of groundwater contamination or remediate the groundwater in the UB. The Unit Operator shall research and develop Projects to mitigate or remove groundwater contaminants in the UB.
B. The Unit Operator shall determine in the Plan the Projects (if any), water charges (if any), or other methods to improve groundwater quality or mitigate the effects of contamination.

14 Projects and Facilities

A. Any existing groundwater facilities listed in Appendix _____ (number) shall be deemed to be Projects and the Parties holding the existing rights in such property shall transfer their rights therein to the Parties collectively in proportion to their number of Shares.
B. Except those properties listed in Appendix _____ (number), each Shareholder shall retain its rights to all other facilities, wells and other real property and tangible personal property acquired prior to the Effective Date, and no such property shall be deemed to be Projects.
C. All rights transferred under Section 11(A) shall be deemed "as is" and with no warranty as to merchantability, fitness for a particular purpose, conformity to models or samples of materials, use, maintenance, condition, capacity or capability.

15 Discoveries

A. Should any Project or Exploration Activity discover any commercially viable pore spaces or groundwater sources, the Unit Operator shall promptly notify the Shareholders on which the discovery was made.
B. Shareholders on which the Project or Exploration Activity were discovered shall have no rights with respect to the Project or Exploration Activity that made the discovery except as part of the Decommissioning.

16 Redetermination of Shares

A. If a subsurface Discovery or Exploration Activity results in the discovery of additional pore spaces or groundwater sources, Shares shall be automatically reallocated after the Effective Date.
B. If additional properties are added to the UB, Shares shall be automatically reallocated after the Effective Date.
C. The Unit Operator shall promptly notify all Shareholders of the number of Shares possessed by the Shareholder, the reason for the Redetermination, and the total number of Shares in the UB.
D. Shareholders shall not claim damages or ameliorative damages for any Redetermination of Shares in the UB against the Unit Operator or any Shareholder.

17 Non-Unit Activities

A. Shareholder shall bear all costs and liabilities for Non-Unit activities conducted within the UB on property owned by the Shareholder.
B. Shareholders shall keep the UB and Projects free and clear of all encumbrances of every kind created by or arising from such Non-Unit activity.

18 Decommissioning

A. This Agreement may be terminated by a vote of _____% of the Shareholders and written consent of the Unit Operator.
B. After Decommissioning, the Projects located on Shareholder property are transferred into the possession of the landowner.
C. The Shareholders collectively maintain liability and costs for all well plugging and other costs as a result of Decommissioning.
D. The Unit Operator is held harmless for all activities occurring after the Decommissioning Date.

19 Renewal of the Agreement

A. The renewal procedure is conducted by the Committee.
B. The renewal procedure is conducted as described in Section 4(f) of this Agreement nine (9) years and (6) months after the Effective Date.
C. If a change in Unit Operator occurs during the renewal of this Agreement, the outgoing Unit Operator shall transfer operation, maintenance, and control of all Projects and responsivities borne by the Unit Operator to the incoming Unit Operator ten (10) years from the Effective Date of this Agreement.

Glossary

Ad coelum doctrine Shortened version of the Latin phrase *"cujus est solum ejus est usque ad coelum et usque ad inferos,"* an ancient doctrine that give ownership rights to the surface landowner of the materials below the surface. This doctrine often determines ownership of solid materials in the subsurface.

Aquifer Various definitions in technical and legal literature, but typically an extensive subterranean strata potentially capable of producing usable quantities of water, consisting of a saturated or unsaturated zone. Aquifers may be confined or unconfined.

Aquifer governance A governance approach incorporating the saturated and unsaturated portion of the aquifer, subsurface biology, geothermal, and other aquifer transresources.

Aquifer, confined An aquifer that is separated from other aquifers by a substantially impermeable layer, identified by the pressure at a location being higher than the top of the confined layer.

Aquifer, fossil An aquifer storing ancient, nonrenewable groundwater.

Aquifer, unconfined An aquifer that rests upon a substantially impermeable layer, often located near the surface, constituting the water table.

Collective resource governance The operation, distribution, and control of a collection of transresources by the community, regulators, and public interests.

Collective resource management The operation, distribution, and control of a single resource by the community, regulators, and public interests.

Common property Ownership interest is held by a group of individuals collectively.

Correlative rights A doctrine that determines the ownership or access to subsurface resources. The concept originates from the ad coelum doctrine.

Ferae naturae doctrine An ancient concept that possession occurs when something is captured and brought under an individual's control. This doctrine is used to determine ownership of hydrocarbons, and groundwater in Texas.

Groundwater governance A governance approach focusing on groundwater availability and allocation, only indirectly addressing aquifer storage and other transresources.

Natural resource management The operation, distribution, and control of a fundamental resource through top-down regulation, indirectly influenced by the community and public interests, onto individual holders of resource access rights.

Negative externality   An effect on other users of a resource through individual use of the resource that creates social costs not felt by the individual user. For an aquifer, examples include pressure losses, subsidence, induced recharge from streams, water quality losses, and pollution.

Positive externality   An effect on other users caused by using a resource that creates social benefits not felt by the individual user. For an aquifer, examples include increasing aquifer storage potential by pumping, artificial recharge due to flood irrigation, or improving quality from artificial storage projects.

Prior appropriation   A priority system granting senior rights with stronger claims to access to water resources. The doctrine is derived from the practice of miners and mining claims during the gold rush.

Private law   Areas of the law related to individual conduct to other individuals. Typical areas of private law are contracts, torts, commercial law, property, and rights. The main holders of private rights are individuals.

Private property   Ownership interest in fee simple or privilege to use a resource on an individual-use basis.

Pseudo-common property   Individual private ownership rights are voluntarily coordinated, creating new relationships of rights, while preserving the underlying individual rights and obligations.

Public law   Areas of the law related to general welfare, morals, and social impacts. These laws are often statutory, representing the public interest in general. The main holder of public law is the state.

Public property   Ownership interest is held in a national, state, or local government for the benefit of the public.

Redetermination   The reallocation of equity interests within a unit, typically conducted at certain times over the life of the unitization agreement.

Riparianism   A water law doctrine that allows landowners the right to the use of the natural flow of water adjacent to their property.

Safe or sustainable yield   Balancing recharge with withdrawals from the aquifer. This approach mischaracterizes aquifer hydraulics.

Sustainability   The balancing of economic, social, and environmental needs ensuring the continuation of the system as a whole.

Transresource   A resource within a network of interconnected physical and social resources tied to a fundamental resource, such as human, social, financial, natural, and physical resources. In economics, a related concept is positive or negative externalities. Transresources are physically interconnected resources within an aquifer system.

   For example, aquifers contain groundwater, storage spaces, geothermal heat, cooling properties, contamination natural and human-introduced contaminants, biological components, hydrocarbons and gases, minerals, and chemical processes; each resource is a transresource. For aquifers, groundwater is only one transresource among many.

Unit operator   The organization that manages a unitized resource for the benefit of the transresource community.

Unit plan   The plan created by a unit operator to manage the facilities and development of a common pool for the benefit of the unit.

Unitization   The operation of a series of interconnected resources as a single physical unit, with the goal of maximizing the social and economic benefits, as well as the efficiency of business ecosystems, associated with the use and reuse of the resource as a whole, regardless of the nature of the individual possessory interests in that resource, using a collective rights-based contract.

Usufructuary rights A right that is limited to the use of a resource, but not the actual ownership of the resource. Private water rights are often considered usufructuary rights, while the public owns the water itself.

Bibliography

Agua Caliente's Memorandum in Support of Motion for Summary Judgement on Phase II
Issues, *Agua Caliente Band of Cahuilla Indians* v. *Coachella Valley Water District
et al.*, United States District Court for the Central District of California, Eastern
Division. [Unpublished Court Document] (2018) (no. CV 13-00883-JGB-SPX).

Alley, W. (2007). Another Water Budget Myth: The Significance of Recoverable Ground
Water in Storage. *Ground Water*, 45(3), 251–251.

Anderson, T. L. (1983). *Water Crisis: Ending the Policy Drought*. Baltimore; London:
Johns Hopkins University Press.

Anderson, T. L., and Snyder, P. (1997). *Water Markets: Priming the Invisible Pump*.
Washington, DC: Cato Institute.

Asmus, D. F., and Weaver, J. L. (2006). Unitizing Oil and Gas Fields around the World:
A Comparative Analysis of National Laws and Private Contracts. *Houston Journal of
International Law*, 28(1), 3–197.

Association of International Petroleum Negotiators (AIPN) (2006). Model Unitization and
Unit Operating Agreement. Available at www.aipn.org/model-contracts/

Aubert, A. H., Bauer, R., and Lienert, J. (2018). A Review of Water-Related Serious
Games to Specify Use in Environmental Multi Criteria Decision Analysis.
Environmental Modelling & Software, 105, 64–78.

Barbanell, E. (2001). *Common-Property Arrangements and Scarce Resources: Water in
the American West*. Westport, CT: Praeger.

Beggs, D., and Stockdale, J. (2014). Unitisation and Unitisation Agreements. In G. Picton-
Turbervill, ed., *Oil and Gas: A Practical Handbook*, 2nd ed. London: Globe Law and
Business.

Birks, D., Whittall, S., Savill, I., Younger, P. L., and Parkin, G. (2013). Groundwater
Cooling of a Large Building Using a Shallow Alluvial Aquifer in Central London.
Quarterly Journal of Engineering Geology and Hydrology, 46, 189–202.

Blomquist, W. A. (1992). *Dividing the Waters: Governing Groundwater in Southern
California*. San Francisco: ICS Press.

Blumberg, H., and Collins, G. (2016). Implementing Three-Dimensional Groundwater
Management in a Texas Groundwater Conservation District. *Texas Water Journal*,
7(1), 69–81.

Bonini, A. (2018). The Hammer and the Hand: Pluralistic Groundwater Governance and
Conflict Transformation in Oregon's Malheur Lake Basin. Unpublished M.S. Thesis,
University of Oregon, Eugene.

Bredehoeft, J. (1997). Safe Yield and the Water Budget Myth. *Ground Water*, 35(6), 929–929.

Bredehoeft, J. (2005). The Conceptualization Model Problem – Surprise. *Hydrogeology Journal*, 13(1), 37–46.

Bredehoeft, J. (2007). It Is the Discharge. *Ground Water*, 45(5), 523–523.

Brough, R. (2019). Summit Water Distribution Company Planning for the Future, Radio Interview and Transcript, KPCW. Available at www.kpcw.org/post/summit-water-distri bution-company-planning-future#stream/0

Brown, J. A. (2015). Uncertainty Below: A Deeper Look into California's Groundwater Law. *Environs: Environmental Law and Policy Journal*, 39(1), 45–95. Available at https://environs.law.ucdavis.edu/volumes/39/1/Articles/Brown-Macroed.pdf

Buckley, S. E. (1951). *Petroleum Conservation* [Henry L. Doherty memorial volume]. New York: American Institute of Mining and Metallurgical Engineers.

Burchi, S. (2018). Legal Frameworks for the Governance of International Transboundary Aquifers: Pre- and Post-ISARM Experience. *Journal of Hydrology: Regional Studies*, 20, 15–20.

California Division of Water Resources. (2016). Modeling Best Management Practice No. 5 for the Sustainable Management of Groundwater, 41 pp. Available at https://water.ca.gov/Programs/Groundwater-Management/SGMA-Groundwater-Management/Best-Management-Practices-and-Guidance-Documents

Campbell, R. J. (1925). Man of Many Millions Lacks Roof Over Head in Heart of Biggest City. *The Brooklyn Daily Eagle* (Brooklyn, NY), May 24, p. 99.

Carlson, J. (2011). A Critical Resource or Just a Wishing Well? A Proposal to Codify the Law on Transboundary Aquifers and Establish an Explicit Human Right to Water. *American University International Law Review*, 26(5), 1409–1436.

Carlton, J. B. (2007). Special Districts, Antitrust Law & the Infrastructure Crisis. *Gonzaga Law Review*, 43(3), 671–700.

Cavallo, A. (2004). Hubbert's Petroleum Production Model: An Evaluation and Implications for World Oil Production Forecasts. *Natural Resources Research*, 13(4), 211–221.

Chaffin, B. C., Craig, R. K., and Gosnell, H. (2014). Resilience, Adaptation, and Transformation in the Klamath River Basin Social-Ecological System. *Idaho Law Review*, 51(1), 157–193.

Chamberlin, T. C. (1897). The Method of Multiple Working Hypotheses. *Journal of Geology*, 5, 837–848.

Closas, A., and Villholth, K. G. (2016). *Aquifer Contracts: A Means to Solving Groundwater Over-Exploitation in Morocco?* Colombo, Sri Lanka: International Water Management Institute. Groundwater Solutions Initiative for Policy and Practice Case Profile Series No. 01. Available at http://gripp.iwmi.org/gripp/publications/case-profile-series/issue-01.pdf

Clyde, S. E. (2011). Beneficial Use in Times of Shortage: Respecting Historic Water Rights While Encouraging Efficient Use and Conservation. *The Water Report*, 83, 1–13.

Cobourn, K. (2011). Dynamic Feedback between Surface and Groundwater Systems: Implications for Conjunctive Management. *Agricultural and Applied Economics Association, 2011 Annual Meeting, July 24–26, 2011, Pittsburgh, PA.*

Cody, K., Smith, S., Cox, M., and Andersson, K. (2015). Emergence of Collective Action in a Groundwater Commons: Irrigators in the San Luis Valley of Colorado. *Society & Natural Resources*, 28(4), 1–18.

Cosens, B., Craig, R., Hirsch, S. et al. (2017). The Role of Law in Adaptive Governance. *Ecology and Society*, 22(1), 1–30.

Craig, R., and Ruhl, J. (2014). Designing Administrative Law for Adaptive Management. *Vanderbilt Law Review*, 67(1), 1–89.

Cuthbert, M. O., Gleeson, T., Moosdorf, N. et al. (2019). Global Patterns and Dynamics of Climate–Groundwater Interactions. *Nature Climate Change.* DOI: https://doi.org/ 10.1038/s41558-018-0386-4

Daniels, S., and Walker, G. B. (2001). *Working through Environmental Conflict: The Collaborative Learning Approach.* Westport, CT: Praeger.

David, M. (1996). Exploration, Appraisal and Development Farmout Agreements. In M. David, ed., *Upstream Oil and Gas Agreements: With Precedents.* London: Sweet & Maxwell.

Doherty, H. L. (1924). Letter from Industrialist Henry L. Doherty to the President of the United States Calvin Coolidge, August 11, 1924 [Letter], in Hardwicke, R. E. (1948). *Antitrust Laws, et al.* v. *Unit Operation of Oil and Gas Pools.* New York: American Institute of Mining and Metallurgical Engineers.

Driscoll, F. G. (1986). *Groundwater and Wells*, 2nd ed. St. Paul, MN: Johnson Division.

Easo, J. (2014). Petroleum Contracts: Licenses, Concessions, Production-Sharing Agreements and Service Contracts. In G. Picton-Turbervill, ed., *Oil and Gas: A Practical Handbook,* 2nd ed. London: Globe Law and Business.

Eckman, D. W. (1973). Statutory Fieldwide Oil and Gas Units: A Review for Future Agreements. *Natural Resources Lawyer*, 6(3), 339, 339–387.

Eckstein, G. (2017). *The International Law of Transboundary Groundwater Resources.* Oxford: Routledge.

Embleton, D. (2012). Use of Exempt Wells as Natural Underground Storage and Recovery Systems. *Journal of Contemporary Water Research & Education*, 148(1), 44–54.

English, W. (1996). Unitisation Agreements. In M. David, ed., *Upstream Oil and Gas Agreements: With Precedents*. London: Sweet & Maxwell.

Environmental Defense Fund. (2019). The Groundwater Game. Available at www.edf.org/ ecosystems/groundwater-game

EPA-Environmental Protection Agency. (2013). Introduction to In Situ Bioremediation of Groundwater. Available at www.epa.gov/sites/production/files/2015-04/documents/ introductiontoinsitubioremediationofgroundwater_dec2013.pdf

Essaid, H., and Caldwell, R. (2017). Evaluating the Impact of Irrigation on Surface Water – Groundwater Interaction and Stream Temperature in an Agricultural Watershed. *Science of the Total Environment*, 599–600, 581–596.

Faigman, D. (1999). *Legal Alchemy: The Use and Misuse of Science in the Law*. New York: W. H. Freeman.

Fennell, L. A. (2011). Ostrom's Law: Property Rights in the Commons. *International Journal of the Commons* 5(March 1), 9–27.

Fisher, R., Ury, W., and Patton, B. (2011). *Getting to Yes: Negotiating Agreement without Giving In*. New York: Penguin Books.

Flatt, V. B. (2009). Paving the Legal Path for Carbon Sequestration from Coal. *Duke Environmental Law & Policy Forum*, 19, 211–246.

Food and Agricultural Organization of the United Nations-FAO. (2006). *FAO Training Manual for International Watercourses/River Basins Including Law, Negotiation, Conflict Resolution, and Simulation Training Exercises*. Rome: FAO.

Foster, S. R. (2011). Collective Action and the Urban Commons. *Notre Dame Law Review*, 87(1), 57–134.

Foster, S. R., and Ait-Kadi, M. (2012). Integrated Water Resources Management (IWRM): How Does Groundwater Fit In? *Hydrogeology Journal*, 20, 415–418.

Foster, S., Garduno, H., Tuinhof, A., and Tovey, C. (2009). Groundwater Governance – Conceptual Framework for Assessment of Provisions and Needs. World Bank GW MATE Strategic Overview Series (1), Washington, DC.

Foster, S., Nanni, M., Kemper, K., Garduno, H., and Tuinhof, A. (2003). Utilization of Non-Renewable Groundwater: A Socially-Sustainable Approach to Resource Management. GW Mate Briefing Note Series No. 11.

Freeze, R., and Cherry, J. (1979). *Groundwater.* Englewood Cliffs, NJ: Prentice-Hall.

Getches, D., Zellmer, S. B., and Amos, A. L. (2015). *Water Law in a Nutshell.* St. Paul, MN: West Academic.

Getzler, J. (2004). *A History of Water Rights at Common Law.* Oxford: Oxford University Press.

Ghosh, S., and Willett, K. (2012). Property Rights Revisited: An Analytical Framework for Groundwater Permit Market under Rule of Capture. *Journal of Natural Resources Policy Research*, 4(3), 143–159.

Gleick, P. H. and Palaniappan, M. (2010). Peak Water Limits to Freshwater Withdrawal and Use. *Proceedings of the National Academy of Sciences,* 107(25), 11155–11162.

Greater Harney Valley Groundwater Study Advisory Committee. (2017). April 20, 2017 Meeting Minutes. Available at https://apps.wrd.state.or.us/apps/misc/vault/vault .aspx?Type=WrdNotice&notice_item_id=7939

Greater Harney Valley Groundwater Study Advisory Committee. (2018). Greater Harney County Groundwater Advisory Committee Charter. Available at www.oregon.gov/ OWRD/programs/GWWL/GW/HarneyBasinStudy/Documents/GWSAC_Charter_ 2018JUL17_FINAL_v.2.pdf

Griggs, B. (2014). *Lessons from Kansas: A More Sustainable Groundwater Management Approach.* Stanford: Water in the West. Available at http://waterinthewest.stanford.edu/ news-events/news-press-releases/lessons-kansas-more-sustainable-groundwater-manage ment-approach

van der Gun, J., and Custodio, E. (2017). Governing Extractable Subsurface Resources and Subsurface Space. In K. G. Villholth, E. López-Gunn, K. Conti, A. Garrido, and J. van der Gun, eds., *Advances in Groundwater Governance.* London: CRC Press.

Hagerty, C. L., and Uzel, J. C. (2013). Proposed U.S.-Mexico Transboundary Hydrocarbons Agreement: Background and Issues for Congress: Congressional Research Service, 7-5700. Available at www.crs.gov

Hansen, J. (2010). It Takes a District: Utah Landowners Control Groundwater Use Escalante Valley Citizens Plan to Save Their Declining Aquifer. *High Country News,* May 10, 2010. Available at www.hcn.org/issues/42.8/it-takes-a-district

Hardin, G. (1968). The Tragedy of the Commons. *Science,* 162(3859), 1243–1248.

Hardwicke, R. E. (1948). *Antitrust Laws, et al.* v. *Unit Operation of Oil and Gas Pools.* New York: American Institute of Mining and Metallurgical Engineers.

Hayton, R., and Utton, A. (1989). Transboundary Groundwaters; the Bellagio Draft Treaty. *Natural Resources Journal*, 29(3), 663–722.

Hepburn, S. J. (2014). Ownership Models for Geological Sequestration: A Comparison of the Emergent Regulatory Models in Australia and the United States. *Environmental Law Reporter*, 44(4), 10310–10325.

Hockaday, S., Jarvis, T., and Taha, F. (2017). Serious Gaming in Water, mediate.com. [Online] Available at www.mediate.com/articles/HockadayS1.cfm

Horwitz, M. J. (1982). The History of the Public/Private Distinction. *University of Pennsylvania Law Review*, 130(6), 1423–1428.

House, K. (2016). Valley Caught in the Middle. *Oregonian*, August 26, 2016. Available at www.oregonlive.com/environment/index.ssf/page/draining_oregon_day_2.html

House, K., and Graves, M. (2016). Draining Oregon. *Oregonian*, August 26, 2016. Available at http://media.oregonlive.com/environment_impact/other/Draining_Oregon_0826d.pdf

Igiehon, M. O. (2014). Decommissioning of Upstream Oil and Gas Facilities. In G. Picton-Turbervill, ed., *Oil and Gas: A Practical Handbook,* 2nd ed. London: Globe Law and Business.

International Groundwater Resources Assessment Center (n.d.). *The Groundwater Game: A Serious Game on Improving Groundwater Management through Cooperation and Collective Action.* Delft, the Netherlands: UN-IGRAC.

International Law Commission. (2008). Report of the International Law Commission Sixteenth Session. Available at http://legal.un.org/docs/?symbol=A/63/10(SUPP).

Isaak, M. T. (2012). Tragedy of the Commons Game. *Arizona Water Resource Newsletter* (Winter). [Online] Available at https://wrrc.arizona.edu/awr/w12/commonsgame

Jarvis, W. T. (2011). Unitization: A Lesson in Collective Action from the Oil Industry for Aquifer Governance, Theme Issue on Transboundary Groundwater. *Water International*, 36(5), 611–622.

Jarvis, W. T. (2014). *Contesting Hidden Waters: Conflict Resolution for Groundwater and Aquifers.* Earthscan Water Text Series. London; New York: Routledge, Taylor & Francis Group.

Jarvis, W. T. (2018a). Cooperation and Conflict Resolution in Groundwater and Aquifer Management. In K. G. Villholth, E. López-Gunn, K. Conti, A. Garrido, and J. van der Gun, eds., *Advances in Groundwater Governance.* London: CRC Press.

Jarvis, W. T. (2018b). Scientific Mediation through Serious Gaming Facilitates Transboundary Groundwater Cooperation. *Water Resources IMPACT*, 20(3), 21–22.

Kaffine, D., and Costello, C. (2011). Unitization of Spatially Connected Renewable Resources. *The B.E. Journal of Economic Analysis & Policy*, 11(1), https://doi.org/10.2202/1935-1682.2714.

Kangas, A., Rasinmäki, J., Eyvindson, K., and Chambers, P. (2015). A Mobile Phone Application for the Collection of Opinion Data for Forest Planning Purposes. *Environmental Management*, 55(4), 961–971.

Keats, A., and Tu, C. (2016). Not All Water Stored Underground Is Groundwater: Aquifer Privatization and California's 2014 Groundwater Sustainable Management Act. *Golden Gate University Environmental Law Journal*, 9(1), 93–108.

Keiter, R., and Ruple, J. (2011). *Topical Report: Conjunctive Surface and Groundwater Management in Utah: Implications for Oil Shale and Oil Sands Development.* Salt Lake City: University of Utah, Institute for Clean and Secure Energy. DOE Award No. DE-FE0001243. 77 pp.

Kendy, E., and Bredehoeft, J. (2006). Transient Effects of Groundwater Pumping and Surface-Water-Irrigation Returns on Streamflow. *Water Resources Research*, 42(8), DOI: https://doi.org/10.1029/2005WR004792.

Klein, H. E. (2008). An Investigation of Groundwater Recharge and Residence Time in the Snyderville Basin, Summit County, Utah Using Noble Gases, Tritium, and Terrigenic Helium. Unpublished M.S. Thesis, University of Utah, Salt Lake City.

Knowlton, D. R. (1939). Unitization – Its Progress and Future, Drilling and Production Practice. American Petroleum Institute Report, 39, 630–635.

Koda, K. (2007). Analysis of Water Right Transfers and Injury Quantification in a Prior Appropriation System – A Perspective from Actuarial Mathematics. Unpublished Honors B.S. Thesis. Oregon State University, Corvallis.

Kokimova, A. (2019). Developing a Transboundary Groundwater Model in the Water Scarce Region of Central Asia: A Case Study of the Pretashkent Transboundary Aquifer.

M.S. Thesis, IHE Delft Institute for Water Education, Delft, the Netherlands. Available at www.un-igrac.org/resource/developing-transboundary-groundwater-model-water-scarce-region-central-asia-case-study

Kramer, B., and Anderson, O. (2005). The Rule of Capture – An Oil and Gas Perspective. *Environmental Law*, 35(4), 899–954.

Kwaku Kyem, P. (2004). Of Intractable Conflicts and Participatory GIS Applications: The Search for Consensus Amidst Competing Claims and Institutional Demands. *Annals of the Association of American Geographers*, 94(1), 37–57.

La Marche, J. (2017). Climate and Hydrologic Cycles and Trends in PNW Basin & Range: Understanding Changing Patterns of Water Availability. [Presentation to Harney Basin Study Advisory Committee, October 17, 2017]. Available at https://apps.wrd.state.or.us/apps/misc/vault/vault.aspx?Type=WrdNotice&notice_item_id=7956

van Laerhoven, F., and Berge, E. (2011). The 20th Anniversary of Elinor Ostrom's *Governing the Commons*. *International Journal of the Commons*, 5(1), 1–8.

Lai, L., Davies, S., and Lorne, F. (2016). Creation of Property Rights in Planning by Contract and Edict: Beyond "Coasian Bargaining" in Private Planning. *Planning Theory*, 15(4), 418–434.

Lawrence, R. J. (2010). Beyond Disciplinary Confinement to Imaginative Transdisciplinary. In V. A. Brown, J. A. Harris, and J. Y. Russell, eds., *Tackling Wicked Problems through the Transdisciplinary Imagination*. Washington, DC: Earthscan, 16–30.

Leahy, T. C. (2016). Desperate Times Call for Sensible Measures: The Making of the California Sustainable Groundwater Management Act. *Golden Gate University Environmental Law Journal*, 9(1), 5–36.

Leonard, A. (1970). Ground-Water Resources in Harney Valley, Harney County, Oregon. State of Oregon. Available at www.oregon.gov/owrd/docs/Place/Malheur_Lake_Basin/USGS_OWRD_GW_Report_16_Harney_Leonard_1970.pdf

Libecap, G. (2005). The Problem of Water. Essay prepared National Bureau of Economic Research. Hoover Institution. Available at http://citeseerx.ist.psu.edu/viewdoc/summary?doi=10.1.1.61.3872

Libecap, G. (2007). The Assignment of Property Rights on the Western Frontier: Lessons for Contemporary Environmental and Resource Policy. *The Journal of Economic History*, 67(2), 257–291.

Libecap, G. (2008). Open-Access Losses and Delay in the Assignment of Property Rights. *Arizona Law Review*, 50(2), 379–408.

Lindholm, G. F. (1996). Summary of the Snake River Plain Regional Aquifer-System Analysis in Idaho and Eastern Oregon, U.S. Geological Survey Professional Paper 1408-A.

Linton, J., and Brooks, D. (2011). Governance of Transboundary Aquifers: New Challenges and New Opportunities. *Water International*, 36(5), 606–618.

Linton, J., and Budds, J. (2014). The Hydrosocial Cycle: Defining and Mobilizing a Relational-Dialectical Approach to Water. *Geoforum*, 57, 170–180.

Lowe, J. (2014). *Oil and Gas Law in a Nutshell. Nutshell Series*. St. Paul, MN: West Academic.

MacDougal, D. (2016). Consensus and Conflict in Oregon's Troubled Waters – A Tale of Four Basins. Marten Law Newsletters (posted February 22, 2016).

Margat, J., and van der Gun, J. (2013). *Groundwater around the World: A Geographic Synopsis*. Leiden, the Netherlands: CRC Press/Balkema.

Martin-Nagle, R. (2016). *Transboundary Offshore Aquifers: A Search for a Governance Regime*. Leiden, the Netherlands: Brill Research Perspectives/Brill.

Martin-Nagle, R. (2020). *Governance of Offshore Freshwater Resources.* Leiden, the Netherlands: Brill-Nijhoff.

Max-Neef, M. A. (2005). Foundations of Transdisciplinary. *Ecological Economics,* 53(1), 5–16.

McIntyre, O. (2011). International Water Resources Law and the International Law Commission Draft Articles on Transboundary Aquifers: A Missed Opportunity for Cross-Fertilization? *International Community Law Review,* 13(3), 237–254.

Mekonnen, A., and Gorsevski, P. (2015). A Web-Based Participatory GIS (PGIS) for Offshore Wind Farm Suitability within Lake Erie, Ohio. *Renewable and Sustainable Energy Reviews,* 41(C), 162–177.

Mid-Continent Oil and Gas Association. (1930). *Handbook on Unitization of Oil Pools.* St. Louis, MO: Blackwell Wielandy.

Miller, S. (2017). Overdrafting Oregon: The Case against Groundwater Mining. *Environmental Law,* 47(2), 519–540.

Moffitt, S. (2012). Summit County Declared as Drought Disaster Area. *The Park Record,* July 17, 2012. Available at www.parkrecord.com/news/summit-county-declared-as-drought-disaster-area/

Mondo, H. (2018). A Look Beneath the Surface: Developing a Transboundary Groundwater Governance Framework and Agreement for the Memphis Sand Aquifer. Unpublished M.S. Thesis, Oregon State University, Corvallis.

Moore, C., and Jarvis, T. (2020). Scientific Mediation and Serious Gaming: New Models for Dealing with the Old Problem of Dueling Experts. *Rocky Mountain Mineral Law Foundation Journal,* 57(1), 35–49.

Moore, C., Jarvis, T., and Wentworth, A. (2015). Scientific Mediation, mediate.com. www.mediate.com/articles/JarvisT1.cfm

Mukherji, A., and Shah, T. (2005). Groundwater Socio-Ecology and Governance: A Review of Institutions and Policies in Selected Countries. *Hydrogeology Journal,* 13(1), 328–345.

Newman, A. (2012). *Edwards Aquifer Authority* v. *Day* and the Future of Groundwater Regulation in Texas. *Review of Litigation,* 31(2), 403–434.

New York Times (1939). Henry L. Doherty, Oil Man Dies at 69. *New York Times,* December 27, 1939.

Nyerges, T., Jankowski, P., Tuthill, D., and Ramsey, K. (2006). Collaborative Water Resource Decision Support: Results of a Field Experiment. *Annals of the Association of American Geographers,* 96(4), 699–725.

Olien, R., and Olien, D. (2002). *Oil in Texas: The Gusher Age, 1895–1945.* Clifton and Shirley Caldwell Texas Heritage Series. Austin: University of Texas Press.

Onorato, W. (1977). Apportionment of an International Common Petroleum Deposit. *International and Comparative Law Quarterly,* 26(2), 324–337.

Oregon Water Resources Department. (2016). Water Rights in Oregon: An Introduction to Oregon's Water Laws. Available at www.oregon.gov/owrd/PUBS/docs/aquabook.pdf

Ostrom, E. (1965). Public Entrepreneurship: A Case Study in Ground Water Basin Management. Unpublished Ph.D. Dissertation, University of California, Los Angeles.

Ostrom, E. (1990). *Governing the Commons: The Evolution of Institutions for Collective Action.* Cambridge: Cambridge University Press.

Ostrom, E., Cox, M., and Schlager, E. (2014). An Assessment of the Institutional Analysis and Development Framework and Introduction of the Social-Ecological Systems Framework. In P. Sabatier and C. M. Weible, eds., *Theories of the Policy Process,* 3rd ed. Boulder, CO: Westview Press/Perseus Books Group.

Oswald, L. D. (1995). New Directions in Joint and Several Liability under CERCLA? *University of California Davis Law Review,* 28(2), 299–365.

Owen, D. (2013). Taking Groundwater. *Washington University Law Review*, 91(2), 253–307. Available at https://openscholarship.wustl.edu/cgi/viewcontent.cgi?article=6043&context=law_lawreview

Pagel, M. (2011). New Water Management Model. *The Water Report* (87), 1–6.

Pagel, M. (2016). Oregon's Umatilla Basin Aquifer Recharge and Basalt Bank. *AMPInsights*, The Rock Report. https://static1.squarespace.com/static/56d1e36d59827e6585c0b336/t/5805466815d5dbb1ab59a238/1476740731982/Oregon-Groundwater-Pagel.pdf

Peck, J., Illigner, R., Wiley, J., and Owen, C. (2019). Groundwater Management: The Movement toward Local, Community-Based, Voluntary Programs. *Kansas Journal of Law and Public Policy*, 29(1), 1–49.

Peredo, A. M., Haugh, H. M., and McLean, M. (2017). Common Property: Uncommon Forms of Prosocial Organizing. *Journal of Business Venturing*, 33(5), 591–602.

Perkowski, M. (2017). Judge: Oregon Regulators Properly Shut Down Klamath Wells. *The Capital Press*. Available at www.capitalpress.com/Oregon/20170906/judge-oregon-regulators-properly-shut-down-klamath-wells

Perkowski, M. (2018). Oregon Agency May Be Awash in Red Ink from Water Litigation. *The Capital Press*. Available at www.capitalpress.com/Water/20180319/oregon-agency-may-be-awash-in-red-ink-for-water-litigation-budget?utm_source=Capital+Press&utm_campaign=593b886967-EMAIL_CAMPAIGN_2018_03_20&utm_medium=email&utm_term=0_3bfe2c1612–593b886967–240236530

Perrone, D., and Jasechko, S. (2019). Deeper Well Drilling an Unsustainable Stopgap to Groundwater Depletion. *Nature Sustainability*, 2, 773–782.

Pinchot, G. (1910). *The Fight for Conservation.* Seattle and London: University of Washington Press. 150 pp.

Preacher, A. (2015). Harney County Groundwater: No New Ag Wells. Oregon Public Broadcasting. Available at www.opb.org/news/article/harney-county-water-woes-no-new-groundwater-wells-/

Provencher, B. (1993). A Private Property Rights Regime to Replenish a Groundwater Aquifer. *Land Economics*, 4, 325–340.

Provencher, B., and Burt, O. (1994). A Private Property Rights Regime for the Commons: The Case of Groundwater. *American Journal of Agricultural Economics*, 76(4), 875–888.

Puri, S., and Villholth, K. G. (2017). Governance and Management of Transboundary Aquifers. In K. G. Villholth, E. López-Gunn, K. Conti, A. Garrido, and J. van der Gun, eds., *Advances in Groundwater Governance*. London: CRC Press.

Rivera, A. (2019). What Is the Future of Groundwater? *Groundwater*, 57(5), 661–662.

Robinson, J., Jarvis, W. T., and Tullos, D. (2017). Domestic Well Aquifer Storage and Recovery Using Seasonal Springs. *Water Resources Impact*, 19(5), 22–24.

Rudestam, K., and Langridge, R. (2014). Sustainable Yield in Theory and Practice: Bridging Scientific and Mainstream Vernacular. *Groundwater*, 52, 90–99.

Santelmann, M., McDonnell, J., Bolte, J. Chan, S., Morzillo, A. T., and Hulse, D. (2012). Willamette Water 2100: River Basins as Complex Social-Ecological Systems. *WIT Transactions on Ecology and the Environment*, 155(1), 575–586. Available at www.witpress.com/Secure/elibrary/papers/SC12/SC12048FU1.pdf

Sato, S., and Crocker, T. D. (1977). Property Rights to Geothermal Resources. *Ecology Law Quarterly*, 6(3), 481–569.

Schlager, E. (2004). Common-Pool Resource Theory. In R. F. Durant, D. J. Fiorino, and R. O'Leary, eds., *Environmental Governance Reconsidered: Challenges, Choices, and Opportunities*. Cambridge, MA: The MIT Press, 145–175.

Schroeder, L. (2016). Innovative Partnerships: An Answer to the Tragedy of the Commons. *2nd World Irrigation Forum Papers*, 1(3). Available at www.icid.org/wif2_full_papers/wif2_w.1.3.02.pdf

Schroeder, L., Ure, T., and Liljefelt, S. (2011). Domestic Groundwater Right Exemptions: Competing Uses Put Pressure on Western Water Right Requirements, but Constitutional Right to Life May Trump Prior Appropriation Doctrine. *Willamette Law Review*, 47(3), 405–424.

Schuerhoff, M., Weikard, H. P., and Zetland, D. (2013). The Life and Death of Dutch Groundwater Tax. *Water Policy*, 15(6), 1064–1077.

Scott, I. (2016). Antitrust and Socially Responsible Collaboration: A Chilling Combination? *American Business Law Journal*, 53(1), 97–144.

Shah, T. (2009). *Taming the Anarchy: Groundwater Governance in South Asia.* Washington, DC: Resources for the Future Press.

Shaw, S. (1996). Joint Operating Agreements. In M. David, ed., *Upstream Oil and Gas Agreements: With Precedents*. London: Sweet & Maxwell.

Siepman, S. (2019). Developing a Transboundary Groundwater Model in Central Asia. *International Groundwater Assessment Centre Newsletter*. Available at www.un-igrac.org/stories/developing-transboundary-groundwater-model-central-asia

Singh, K. (2016). Managing Water for Sustainable Development: An Economist's Perspective. *IIM Kozhikode Society & Management Review,* 5(1), 1–4.

Smith, P. (1986). Coercion and Groundwater Management: Three Case Studies and a "Market" Approach. *Environmental Law*, 16(4), 797–882.

Smith, S. M., Andersson, K., Cody, K., Cox, M., and Ficklin, D. (2017). Responding to a Groundwater Crisis: The Effects of Self-Imposed Economic Incentives. *Journal of the Association of Environmental and Resource Economists*, 4(4), 985–990.

Smitherman, L. (2015). Forensic Hydrogeography: Assessing Groundwater Arsenic Concentrations and Testing Methods within the Harney Basin, Oregon. Unpublished M.S. Thesis, Oregon State University, Corvallis. Available at http://ir.library.oregonstate.edu/concern/graduate_thesis_or_dissertations/hh63sz81j

Sprankling, J. (2008). Owning the Center of the Earth. *UCLA Law Review*, 55, 979, 1004.

State of Utah Department of Natural Resources. (2012). Beryl Enterprise Groundwater Management Plan. Available at www.waterrights.utah.gov/groundwater/ManagementReports/BerylEnt/BerylEnterprise_Management_Plan.pdf

Strachan, A. (2001). Concurrency Laws: Water as a Land-Use Regulation. *Journal of Land, Resources, and Environmental Law*, 21, 435–460.

Susskind, L. (2012). Learning from Games: The Debate over Role-Play Simulations. Available at http://theconsensusbuildingapproach.blogspot.com/

Taleb, N. N. (2007). *The Black Swan: The Impact of the Highly Improbable*. New York: Random House.

Tuthill, D.R., Jr., and Carlson, R. D. (2018). Incentivized Managed Aquifer Recharge Basin Scale Implementation Provides Water for Private Users, Groundwater Districts, Municipalities and Others. *The Water Report,* 176, 11–20.

Villholth, K. G., and Conti, K. (2017). Groundwater Governance: Rationale, Definition, Current State and Heuristic Framework. In K. G. Villholth, E. López-Gunn, K. Conti, A. Garrido, and J. van der Gun, eds., *Advances in Groundwater Governance*. London: CRC Press.

Villholth, K. G., Lopez-Gunn, E., Conti, K., Garrido, A., and Van Der Gun, J., eds. (2018). *Advances in Groundwater Governance*. London: CRC Press.

Walker, G. (2011). The Partnership for Coastal Watersheds Collaboration Compact. Available at www.partnershipforcoastalwatersheds.org/

Weaver, B. D. (n.d.). Doherty, Henry Latham. *The Encyclopedia of Oklahoma History and Culture*. Available at www.okhistory.org/publications/enc/entry.php?entry=DO006.

Weaver, J. L. (1986). *Unitization of Oil and Gas Fields in Texas – A Study of Legislative, Administrative, and Judicial Policies*. Washington, DC: Resources for the Future Press.

Weiser, M. (2018). Inside the Ambitious Plan to Replenish a Depleted Aquifer. *Pacific Standard Magazine*, April 12, 2018. Available at https://psmag.com/environment/recharging-a-depleted-aquifer

Wiley, J. S. (2018). Collective Aquifer Governance: It's the Water, and a "Hole" Lot More. [Capstone Project] Oregon State University. Available at https://ir.library.oregonstate .edu/concern/graduate_projects/pn89dd30b?locale=en

Wiley, J. S. (2019). Aquifer Unitization: Applying Collective Governance Agreements to Aquifers. *United States Committee on Irrigation and Drainage Conference Papers*. Available by request from author.

Wolf, A. (2007). Shared Waters: Conflict and Cooperation. *Annual Review of Environmental Resources*, 32(1), 241–269.

Workman, J. G. (2012). Elinor Ostrom: The Patron Saint of Enviropreneurs, *PERC*, 30(2, Fall). Available at www.perc.org/2012/09/14/elinor-ostrom-the-patron-saint-of-enviropreneurs/

Worthington, P. F. (2011). Contemporary Challenges in Unitization and Equity Redetermination of Petroleum Accumulations, *SPE Economics & Management*, 3(1), 10–17.

Worthington, P. F. (2016). Provision for Expert Determination in the Unitization of Straddling Petroleum Accumulations. *The Journal of World Energy Law & Business*, 9(August 4), 254–268. https://doi.org/10.1093/jwelb/jww014

Worthington, P. F. (2018). Petroleum Unitisation: How to Do Better, Transcript of Presentation to the Society of Petroleum Engineers, London Section, on April 24, 2018. *SPE Review*, 31, 4–6, www.spe-london.org/wp-content/uploads/2018/05/SPE-Review-London_May-2018.pdf

Young, M. D. (2014). Designing Water Abstraction Regimes for an Ever-Changing and Ever-Varying Future. *Agricultural Water Management*, 145, 32–38.

Young, M. (2015). Unbundling Water Rights: A Blueprint for Development of Robust Water Allocation Systems in the Western United States. NI R 15-01. Durham, NC: Duke University. Available at http://nicholasinstitute.duke.edu/publications

Zerrenner, A., and Gulley, R. (2016). The Edwards Aquifer. [Online] The Program for the Advancement of Research on Conflict and Collaboration (PARCC), Maxwell School of Citizenship and Public Affairs, Syracuse University. Available at www.maxwell.syr .edu/parcc/eparcc/cases/The_Edwards_Aquifer/

Zhao, Z. J., and Anand, J. (2013). Beyond Boundary Spanners: The "Collective Bridge" as an Efficient Interunit Structure for Transferring Collective Knowledge. *Strategic Management Journal*, 34(13), 1513–1530.

Index

Printed in the United States
by Baker & Taylor Publisher Services